VEDANTA PHILOSOPHY

VEDANTA PHILOSOPHY

Editor-in-Chief

Swami Chidatman Jee Maharaj

ANMOL PUBLICATIONS PVT. LTD.

NEW DELHI - 110 002 (INDIA)

ANMOL PUBLICATIONS PVT. LTD.
H.O.: 4374/4B, Ansari Road, Daryaganj,
New Delhi-110 002 (India)
Ph.: 23278000, 23261597
B.O.: No. 1015, Ist Main Road, BSK IIIrd Stage
IIIrd Phase, IIIrd Block,
Bangalore - 560 085 (India)
Visit us at: www.anmolpublications.com

Vedanta Philosophy

First Published, 2009
ISBN 978-81-261-3618-6

PRINTED IN INDIA

Printed at Mehra Offset Press, Delhi.

Contents

Preface vii

1. End of Vedas and Philosophy of Vedanta 1

2. Purva Mimamsa, Advaita and other Schools of Vedanta 121

3. Vedanta—Essentials of the Higher Values of Life 141

4. The Institution of Shankaracharya 221

Bibliography 249

Index 253

Preface

India colourful and vibrant, a land as diverse as its people. A mosaic of faiths, cultures, customs and languages that blend harmoniously to form a composite whole. One of the world's oldest living civilizations—which gave to the world—the concept of zero, the primordial sound Aum..., Yoga and Buddhism.

She is old. She is young. The two diametrically opposite, different yet unique and dynamic faces of India coexist even at this turn of the millennium. With a history of heritage, learning and civilisation dating back to almost 5000 years or perhaps even more, the modern India presents itself as an ideal country. From snow clad mountains to breezy sparkling blue sea side. From arid deserts with shifting sand dunes to lush green country sides. From ancient relics and architectural ruins dating back into antiquity to modern cities, India has it all.

To the north it is bordered by the world's highest mountain chain, where foothill valleys cover the northernmost of the country's maximum states. Further south, plateaus, tropical rain forests and sandy deserts are bordered by palm fringed beaches. Its southernmost tip is the meeting point of three Oceans.

India is a land of staggering contrast with a mingling of the tradition and modernity that is an unique experience to be in India. One will forever remember the stay in this wonderful Homeland-a land truly of mystery and enchantment. Only India can offer such an astonishing variety of contrasts.

She has given birth to numerous spiritual leaders and founders of some of the popular religions of the present day world. India probably has the most religious diversity in any country. It's the birthplace of Hinduism, Buddhism, Sikhism and Jainism and is among the few places to have a resident

Zoroastrian population. India with her veritable treasure trove of culture, mysticism, philosophy, art, music and architecture to name a few. Much of India's classical music is devotional and the North Indian Hindustani and South Indian Carnatic streams are distinct and both have a complex 'raga' framework. The legacy of dance in India is tremendous. The classical dances of India are numerous such as Kathakali of Kerala, Bharatnatyam of Tamil Nadu, Kuchipudi of Andhra Pradesh, Manipuri and Odissi from Orissa are the prominent dance forms in this country that sways to an altogether novel beat. The Yakshagana, nautanki and puppetry are ancient folk forms that live on till date. The earliest specimens of Indian painting are the ones on the walls of the Ajanta Caves dating back to 2nd century BC. The Mughals had a huge impact on Indian art. The influence of Persian art brought placid garden scenes, illustrations from myths, legends and history into Indian art. Word craft, handicrafts, architecture and sculpture all contribute to this rich and varied domain. Indian literature, both in English and in the vernacular, is ever more popular around the world.

The book aims at presenting perspectives of different aspects of Indian culture. It has topics on Indian classical music and dance, painting, ayurveda and yoga which are traditional and indigenous system of physical and mental wellbeing, language, philosophy and excellence achievements by Indian in the different branches of science and social sciences. It also includes our ancient religious and classic literature.

It is hoped that this book will be welcomed by the general readers as also scholars who are interested in the Indian heritage.

—Editors

1

End of Vedas and Philosophy of Vedanta

The Vedanta philosophy is drawn from the Vedas, which are the traditional Hindu scriptures. The Vedas are a diverse collection of Sanskrit writings extending back several thousand years, concerned with the knowledge of spiritual reality.

The root word, "Ved," means to know in the mystical sense: to know spiritual truth. The suffix "anta" means "end" in the sense of termination and, secondly, of purpose. The end of the Vedas is the philosophical section, the "Vedanta," the portion known as the Upanishads.

The Upanishads are part of the Vedas as the New Testament is part of the Bible. They not only occur at the end of the Vedas, but are the culmination towards which the Vedas move. The Upanishadic texts are unique in their endless emphasis on the central Vedantic discovery: The Supreme Being is the Self and *your* self is *that* Self.

Vedanta's basic principles are:

(1) Nothing exists except the Divine Being, or *Brahman*. It pervades, supports, determines, and explains everything. It is called the One. The doctrine of the Spiritual Oneness of Existence follows from this.

(2) The very nature of the Soul is Divine: the Cosmic Self becomes the individual Self.

(3) The goal of the individual, then, is to realize his/her true nature, which is Divine.

(4) The Universality of religious truth: there are many ways to realize God. (Prophets differ in their interpretation of religion as to details, but not in the essentials.)

Vedanta is a scientific approach to religion and religious truth. There is no one individual associated with the origin of Vedantic philosophy, upon which Hinduism is based. Every other creed was founded by an individual. Not Hinduism. Only principles at the beginning; Truth and Universal laws discovered and passed on by the ancient Vedic sages. The main Vedantic texts are: the Upanishads, the *Bhagavad Gita,* and the *Brahma Sutras.* The Vedanta philosophy is also readily found in the teachings of Buddha, Shankara, Ramakrishna-Vivekananda. It is open ended, a vast cathedral of thought encompassing both the personal and impersonal approaches to divine realization.

Each religion thinks it has the truth, and therefore is superior to others. Every religion differs from others, and wherein they differ is not Vedanta, but rather wherein they are alike. Vedanta becomes self evident when any given religion acts in a universal manner, pleasing everyone. If each religion was deprived of those elements in its creed and practice that separate it from other religions, then what remains that is Vedanta. For example: all religions teach the perfection of self-sacrifice as the highest virtue. Vedanta takes its stand on this at the start. So all religions resemble one another in this respect. They stress the virtue of renunciation, service to others, relating to others as though to one's own self, as though we had in common the same life, the same destiny and the same self, which is a simple but adequate rendering of the prime truth of Vedanta.

Vedanta is without creed, dogma, ritual or doctrine. Vedanta is the mysticism of spirituality. It is what is left over when all the differences that separate religions have been eliminated. Therefore, it is the essence of religion, the truth embedded in the heart of every religion. Vedanta is the Godhead that makes every religion divine.

Vedanta is the chief means whereby each religion will have the incentive to practice tolerance and acceptance toward all other faiths. Every religion, every sect, thinks it has found the truth and it alone has found it. A universal disease!

Throughout history it is the same....To this day. Not only Christians, Jews, and Moslems, but Hindus, Buddhists, Sufis, etc.

How to break the hold of this compulsion?

Vedanta shows the way. It teaches that Truth is universal, that it is continually revealed, that the prophets keep coming, that the Truth may be known via many paths... if followed faithfully, and spiritually.

One more golden virtue of the Vedanta philosophy: With the help of Vedanta we can practice our personal devotions, fervent in our faith and dedication, without feeling that we alone can reach the goal, and convinced quite convinced that others, too, may succeed, though on another path.

Vedanta is not a matter of belief but realization. We are not asked to believe but to realize. If there is a God, then we should be able to experience him.

Not faith alone, but function. Not belief, but behaviour.

Vedanta gives freedom to individuals to advance along their own line of development.

Shankaracharya is the most important teacher of the Advaita school of Vedanta, and his commentaries to the Upanishads, the Bhagavad-Gita and the Brahmasutras define the parameters of *Advaita* thought. However, it must be remembered that all Vedanta philosophy really goes back to the Upanishads, and *Shankaracharya* is regarded as a pre-eminent teacher who continued the Upanishadic tradition. The name *Shankaracharya* has become a title for the heads of the numerous Advaita institutions in India today, because of the great respect and fame associated with it.

The philosophy of Advaita, literally non-dualism, is the premier and oldest extant among the Vedanta schools of Indian philosophy. The Upanishadic quest is to understand Brahman, the source of everything, the *Atman*, the Self, and the relationship between Brahman and *Atman*. The Upanishads explore these issues from different angles. The Advaita school teaches a complete essential identity between Brahman and *Atman*. In other vedantic traditions, the essential relationship between *Atman* and Brahman is understood in different ways.

The Advaita Vedanta has been classified into three sections, named *History*, *Philosophers* and *Philosophy*. The "History" section deals with Shankaracharya, the issues involved in reckoning his date, the living Advaita tradition and related topics. Pre-Shankaran Vedanta, Gaudapada, Shankaracharya, his disciples, Mandana Misra and post-Shankaran advaitins are discussed in appropriate pages under the "Philosophers" section. The "Philosophy" section starts with a brief introduction to various schools of Indian philosophy and the source texts of Vedanta, the Upanishads. Philosophical issues in Advaita Vedanta are examined in detail in this section. More pages on different aspects of Advaita Vedanta and its relation to other systems are under construction.

The Supreme Swan: In the background is an artistic rendering of a swan, with the Sanskrit sentence *Brahmaiva satyam*-Brahman is the only Truth. The swan motif is seen in the seals of many Advaita organizations. The figure seen here has been adapted from the official seal of the Sringeri Matha, an ancient and one of the most important centres of Advaita Vedanta in India. The swan is a very popular motif in traditional Hindu symbolism. It can be found in oil-lamps used in temples and at shrines in people's homes.

The swan has a special association with Advaita Vedanta. The swan is called *hamsa* in the sanskrit language. The greatest masters in the Advaita tradition are called *paramahamsas*-the great swans. The word *hamsa* is a variation of *so'ham*: I am He, which constitutes the highest realization. There are other equivalences between the swan and the advaitin, that make the swan a particularly apt symbol for Advaita Vedanta. The swan stays in water, but its feathers remain dry. Similarly, the advaitin lives in the world, yet strives to remain unaffected by life's ups and downs.

In India, the swan is also mythically credited with the ability to separate milk from water. Similarly, the advaitin discriminates the eternal *Atman* from the non-eternal world. The *Atman* that is Brahman is immanent in the world, just like milk is seemingly inseparably mixed with water, but it can never be truly realized without the *nitya-anitya-vastu viveka*-right discrimination between the eternal and ephemeral-that is essential for the advaitin. The swan is thus a symbol for the *lvanmukta*, who is liberated while still alive in this world, by virtue of having realized Brahman.

Vedanta is a school of philosophy within Hinduism. The word *Vedanta* is a compound of *veda* "knowledge" and *anta* "end, conclusion", translating to "the culmination of the *Vedas*". Vedanta is also called *Uttara Mimamsa*, or the 'latter' or 'higher enquiry', and is often paired with Purva Mimamsa, the 'former enquiry'. Purva Mimamsa, usually simply called Mimamsa, deals with explanations of the fire-sacrifices of the Vedic mantras (in the Samhita portion of the Vedas) and Brahmanas, while Vedanta explicates the esoteric teachings of the raryakas (the "forest scriptures"), and the Upanishads, composed from the 6th century BC until modern times.

History: While the traditional Vedic 'Karma kanda', or ritualistic components of religion, continued to be practiced through the Brahmins as meditative and propitiatory rites to guide society to self-knowledge, more Jnana-or knowledge-centered understandings began to emerge. These are mystical streams of Vedic religion that focused on meditation, self-discipline and spiritual connectivity rather than on rituals.

Etymologically, *veda* means "knowledge" and *anta* means "end", so the literal meaning of the term "Vednta" is "the end of knowledge" or "the ultimate knowledge" or "matter appended to the Veda". In earlier writings, Sanskrit 'Vedanta' simply referred to the Upanishads, the most speculative and philosophical of the Vedic texts. However, in the medieval period of Hinduism, the word Vedanta came to mean the school of philosophy that interpreted the Upanishads. Traditional Vedanta considers scriptural evidence, or shabda pramana, as the most authentic means of knowledge, while perception, or pratyakssa, and logical inference, or anumana, are considered to be subordinate (but valid).

Formalization: The systematization of Vedantic ideas into one coherent treatise was undertaken by Badarayana in the Vedanta Sutra, or Brahma Sutra. The cryptic aphorisms of the Vedanta Sutras are open to a variety of interpretations, resulting in the formation of numerous Vedanta schools, each interpreting the texts in its own way and producing its own sub-commentaries claiming to be faithful to the original. Consistent throughout Vedanta, however, is the exhortation that ritual be eschewed in favour of the individual's quest for truth through meditation governed by a loving morality, secure in the knowledge that infinite bliss awaits the seeker. Near all existing sects of Hinduism are directly or indirectly influenced by the thought systems developed by Vedantic thinkers. Hinduism to a great extent owes its survival to the formation of the coherent and logically advanced systems of Vedanta.

In Modern Times: Advaita Vedanta has influenced a number of modern scientists, philosophers and authors. Nikola Tesla was influenced by the teachings of Swami Vivekananda, who propounded Vedanta vigorously in the West in the late nineteenth and early twentieth centuries. Many other Indian teachers followed in due course, including Yogananda and others who came to North America to preach Vedanta and related philosophies. Erwin Schrödinger claimed to have been inspired by Vedanta in his discovery of quantum theory. According to his biographer Walter Moore: "The unity and continuity of Vedanta are reflected in the unity and continuity of wave mechanics. In 1925, the world view of physics was a model of a great machine composed of separable interacting material particles. During the next few years, Schrödinger and Werner Heisenberg and their followers created a universe based on superimposed, inseparable waves of probability amplitudes. This new view would be entirely consistent with the Vedantic concept of All in One."

Fritjof Capra's book The Tao of Physics is one among several that pursues this viewpoint as it investigates the relationship between modern, particularly quantum, physics and the core philosophies of various Eastern religions, including Hinduism, Buddhism and Taoism. Unfortunately, such writings by western authors often run the risk of oversimplifying and ignoring important differences between Eastern religions. For instance, pre-modern Vedantins argued for the existence of an eternal self, or *Atman*, while Buddhism denies it. However, in recent times, the availability of an increasing number of accurate translations of Vedantic works, commentaries by Western scientists like Schrodinger and Capra, and easier access to original texts have made it possible for modern students of Vedanta and physics to overcome the semantic gap arising due to cultural differences and approach their study in a more informed manner.

Among other prominent Western figures who have been influenced by and commented on Vedanta are Max Muller, Aldous Huxley, Christopher Isherwood, Romain Rolland, Alan Watts, Eugene Wigner, Arnold Toynbee, and Will Durant.

Source Texts: All forms of Vedanta are drawn primarily from the Upanishads, a set of philosophical and instructive Vedic scriptures, which deal mainly with forms of meditation. "The Upanishads are commentaries on the Vedas, their putative end and essence, and thus known as Vedanta = 'End of the Veda'. They are considered the fundamental essence of all the Vedas and although they form the backbone of Vedanta, portions of Vedantic thought are also derived from some of the earlier Aranyakas.

The primary philosophy captured in the Upanishads, that of one absolute reality termed as Brahman is the main principle of Vedanta. The sage Badarayana was one of the major proponents of this philosophy and author of the Brahma Sutras based on the Upanishads. The concept of Brahman – the Supreme Spirit or the eternal, self existent, immanent and transcedent Supreme and Ultimate Reality which is the divine ground of all Being-is central to most schools of Vedanta. The concept of God or Ishvara is also there, and the Vedantic sub-schools differ mainly in how they identify God with Brahman.

The contents of the Upanishads are often couched in enigmatic language, which has left them open to various interpretetions. Over a period of time, several scholars have interpreted the writings in Upanishads and other scriptures like Brahma Sutras according to their own understanding and the need of their time. There are a total of six important interpretations of these source texts, out of which, three (Advaita Vedanta, Vishishtadvaita and Dvaita) are prominent, both in India and abroad. These Vedantic schools of thought were founded by Shri Adishankara, Shri Ramanuja and Shri Madhvacharya, respectively. It should be noted, however, that the Indian pre-Shankara Buddhist writer Bhavya in the Madhyamakahrdaya Karika describes the Vedanta philosophy as "Bhedabheda". Proponents of other Vedantic schools continue to write and develop their ideas as well, although their works are not widely known outside of smaller circles of followers in India.

Doctrines: Vedanta is from the Sanskrit meaning "end of the Vedas." This primarily refers to the Upanishads. Vedanta is the most widespread of the six philosophical systems of Hinduism and has contributed most directly to the development of later Hindu thought. There are three main schools of Vedanta. The basic doctrine of Vedanta philosophy is that the *Atman*, self or soul, is identical with Brahman, the All-Soul or Absolute Reality. Brahman is entire and indivisible so one cannot exist separately as a part or an emanation of Brahman. Nothing real exists besides Brahman.

This is the doctrine of the oldest school of Vedanta called Advaita, or non-dualism, which was expounded by the philosopher Shankara. As reality is nondual, the world of phenomena only appears to be real. It is described as maya, illusion, but it does have empirical viability sustained by avidya, ignorance, and only dispelled by Vidya, knowledge, in the form of Jnana, meditative insight. Shankara's personality is paradoxical, for though strongly teaching the Advaita doctrine, he also composed some beautiful and moving hymns. The teachings of Shankara were elaborated by a long line of scholars.

The second most important school is VisistAdvaita, qualified non-dualism, advocated by Ramanuja. This is the doctrine of one God qualified by cit, souls, and acit, matter. God is the only absolute reality and the only object worthy of love and devotion. Matter and souls are equally ultimate and real and are the visesanas, qualities, of God, and are dependent on God as they exist entirely for and within him. Ramanuja successfully joined personal theism with absolutistic philosophy. Lack of faith was more important to him than ignorance in obscuring God. Bhakti, devotion, was essential to gain liberation.

The Dvaita school was founded by Madhva. Reality consists of two principles, God as the one infinite independent principle, and matter and souls, which is finite reality and dependent on God. The material world is essentially real and the distinction between God and jivas, souls, is real and beginningless. There is a fivefold distinction: between God and jivas, between God and insentient objects, between jivas, between jivas and insentient objects, and between insentient objects. Jivas are subject to God and infinite in number with a gradation according to Karma. Moksha, liberation, is attained by the blissful experience of one's pure nature through devotion as a servant of Krishna.

Besides the three main schools, there have been a number of subschools, some of which have split into further groups. The more significant are the traditions of Nimbarka (see Nimavats), Vallabha (see separate entry), and Chaitanya (see Gaufdiya Vaishnavas). Also see entries on Shankara and Dashanamis. Vedantic doctrines have been important in modern Hindu reform movements, such as the Brahmo Samaj (see separate entry), and to many yoga schools. New forms of Vedanta continue to appear, with more recent examples being Radhakrishnan, Sri Aurobindo and the Ramakrishna Mission.

History: Historically, there are three periods of Vedanta. First is the creative period of the Upanishads. The Upanishads were an intellectual and social rebellion against the closed mechanical priesthood supported by the Brahmanas. The oldest Upanishads date from the eighth century BC and are Vedic in being associated with the Brahmanas, texts dealing with the sacred power in the sacrificial ritual.

These early Upanishads are the Brhadaranyaka, Chandogya, Aitareya, Taittiriya, and Kansitaki Upanishads. Of later origin are the Upanishads of a third group, which has only a slight Vedic association or none at all. Many of them are sectarian. Other texts called Upanishads have been written into modern times.The second historical period is that of the systematisation and harmonisation of the Upanishads. The most important text in this is Bandarayanas's Brahma Sutra or Vedanta Sutra.

This was a synthesis of Upanishadic doctrine written early in the Common Era. After the Brahma Sutra came a long period of development. The grammarian Bhartrhari in about the fifth century conceived Brahman as essentially word symbolised by om. Gandapada commented on the *Mandukya* Upanishad and said in his doctrine of non-origination that the world is maya, only an appearance. Buddhist influence can be seen in his work.The third historical period is that of commentaries on the Brahma Sutra and commentaries on those commentaries as well as other Vedantic works. The most outstanding exponent of Vedanta was Shankara. It is believed by some scholars that Shankara was first a follower of Patanjali's yoga system. Gandapada was the main influence on Shankara, and tradition says he was the teacher of Shankara's teacher. Shankara lived in the eighth century.

Many pupils and teachers continued the work of Shankara including Mandava Misra and Padmapada. Others founded subschools such as Vachaspati, who joined the teachings of Shankara and Mandana Misra. Other writers of the school of Shankara include Vidyaramja in the fourteenth century, Prakasananda and Madhusudana in the sixteenth century, and Sadananda Vyasa in the seventeenth century. Ramanuja, the founder of VisistAdvaita Vedanta, lived from 1017 to 1137 and Madhva, the founder of Dvaita Vedanta, lived from 1238 to 1317.

The followers of Shankara opposed and persecuted Madhva, who probably as a result taught that Shankara was a reincarnation of the demon Manimat from the Mahabharata. Vedanta has had an important impact on the West, starting with the visit of Swami Vivekananda to the World's Parliament of Religions at Chicago in 1893. In 1894 he established the Vedanta Society in New York and talked on Vedanta all over America before returning to India to found the Ramakrishna

Mission, which is today the foremost Vedanta—inspired religious organisation.

Symbols: Vedanta in theory does not have a wide use of symbolism, especially Advaita. But in practice there are symbols. The sacred syllable om symbolises Brahman as being essentially word or speech. There are even images of Shankara worshipped by devotees. An example is the one in a small shrine near Srisailam in Andhra Pradesh on the site where Shankara performed penance during his travels around India.

Modern Vedanta-based groups such as the Ramakrishna Mission have evolved their own symbolism. Ramanuja's school is largely followed by the Shri Vaishnavas, in which devotion plays a large part and is involved with temples and the images of Sri and Vishnu in various forms. Madhva established his main temple himself at Udipi, which he consecrated to Krishna.

In this temple he installed an image of Bala Krishna obtained from Dwarka, and the temple still flourishes. The Madhva school also venerates the other avataras, incarnations, of Vishnu, as well as and the five orthodox gods.

Adherents: Nearly all educated Hindus in modern India are adherents of the Vedanta. About three-quarters follow Shankara's Advaita school, and the remainder are divided among the other schools. Shankara set up the monastic orders of the Dashanamis, with a network all over India. Shri Vaishnavas in South India follow Ramanuja's school and the Madhvas are also mainly in South India.There have been and still are many followers of Vedanta in the West. Aldous Huxley, Christopher Isherwood, and others were attracted to the Vedanta Society, established in New York in 1894.

Many schools of yoga have Vedanta as a philosophical base and these are widespread in the West. In India new forms of Vedanta continue to appear and attract adherents. Reform groups such as the Brahmo Samaj are Vedanta inspired. The eclectic thought of Radhakrishnan is based on the Vedanta, but though influential no actual school was founded by him. Sri Aurobindo established a large ashram at Pondicherry. The Ramakrishna Math and Mission is the major Vedantic organisation with centres all over India and the world. The Ramakrishna Vedanta Centre in England produces a magazine Vedanta (formerly called Vedanta for East and West).

The basic problem of Shankara's philosophy is how such pure consciousness appears, in ordinary experience, to be individualized ("my consciousness") and to be of an object ("consciousness of blue"). As he stated it, subject and object are as opposed to each other as light and darkness, yet the properties of one are superimposed on the other. If something is a fact of experience and yet ought not to be so—*i.e.*, is rationally unintelligible—then this must be false.

According to Shankara's theory of error, the false appearance is a positive, presented entity that is characterized neither as existent (because it is sublated when the illusion is corrected) nor as non-existent (because it is presented, given as much as the real is).

The false, therefore, is indescribable either as being or as non-being, it is not a fiction, such as a round square. Shankara thus introduced a new category of the "false" apart from the usual categories of the existent and the nonexistent. The world and finite individuals are false in this sense: they are rationally unintelligible, their reality is not logically deducible from Brahman, and their experience is cancelled with the knowledge of Brahman.

The world and finite selves are not creations of Brahman; they are not real emanations or transformations of it. Brahman is not capable of such transformation or emanation. They are appearances that are superimposed on Brahman because of man's ignorance. This superimposition was sometimes called adhyasa by Shankara and was often identified with avidya. Later writers referred to avidya as the cause of the error. Thus, ignorance came to be regarded as a beginningless, positive something that conceals the nature of reality and projects the false appearances on it. Shankara, however, did distinguish between three senses of being: the merely illusory (pratibhasika), the empirical (Vyavaharika; which has unperceived existence and pragmatic efficacy), and transcendental being of one, indeterminate Brahman.

In his epistemology, Shankara's followers in general accepted the point of view of the Mimamsa of Kumarila's school. Like Kumarila, they accepted six ways of knowing: perception, inference, verbal testimony, comparison, nonperception, and postulation. In general, cognitions are regarded as modifications of the inner sense in which the pure spirit is reflected or as the pure spirit limited by respective mental modifications. The truth of cognitions is regarded as intrinsic to them, and a knowable fact is accepted as true so long as it is not rejected as false.

In perception a sort of identity is achieved between the form of the object and the form of the inner sense; in fact, the inner sense is said to assume the form of the object. In their theory of inference, the Nyaya five-membered syllogism is rejected in favour of a three-membered one. Furthermore, the sort of inference admitted by the Nyaya, in which the major term is universally present, is rejected, because nothing save Brahman has this property according to the system.

Shankara regarded moral life as a necessary preliminary to metaphysical knowledge and thus laid down strict ethical conditions to be fulfilled by one who wants to study Vedanta. For him, however, the highest goal of life is to know the essential identity of his own self with Brahman, and though moral life may indirectly help in purifying the mind and intellect, over an extended period of time knowledge comes from following the long and arduous process whose three major stages are study of the scriptures under appropriate conditions, reflection aimed at removing all possible intellectual doubts about the non-dualistic thesis, and meditation on the identity of *Atman* and Brahman.

Moksha is not, according to Shankara, a perfection to be achieved; it is rather the essential reality of one's own self to be realized through destruction of the ignorance that conceals it. God is how Brahman appears to an ignorant mind that regards the world as real and looks for its creator and ruler. Religious life is sustained by dualistic concepts: the dualism between man and God, between virtue and vice, and between this life and the next. In the state of Moksha, these dualisms are transcended. An important part of Shankara's faith was that Moksha was possible in bodily existence. Because what brings this supreme state is the destruction of ignorance, nothing need happen to the body; it is merely seen for what it really is—an illusory limitation on the spirit.

Shankara's chief direct pupils were Sureshvara, the author of Varttika ("Gloss") on his bhasya and of Naiskarmya-siddhi ("Establishment of the State of Non-Action"), and Padmapada, author of Panchapadika, a commentary on the first five padas, or sections, of the bhasya. These early pupils raised and settled issues that were not systematically discussed by Shankara himself—issues that later divided his followers into two large groups: those who followed the Vivarana (a work written on Padmapada's Panchapadika by one Prakashatman in the 12th century) and those who followed Vachaspati's commentary (known as Bhamati) on Shankara's bhasya.

Among the chief issues that divided Shankara's followers was the question about the locus and object of ignorance. The Bhamati school regarded the individual self as the locus of ignorance and sought to avoid the consequent circularity (arising from the fact that the individual self is itself a product of ignorance) by postulating a beginningless series of such selves and their ignorances. The Vivarana school regarded both the locus and the object of ignorance to be Brahman and sought to avoid the contradiction (arising from the fact that Brahman is said to be of the nature of knowledge) by distinguishing between pure consciousness and valid knowledge (pramajnana).

The latter, a mental modification, destroys ignorance, and the former, far from being opposed to ignorance, manifests ignorance itself, as evidenced by the judgment "I am ignorant." The two schools also differed in their explanations of the finite individual. The Bhamati school regarded the individual as a limitation of Brahman just as the space within the four walls of a room is a limitation of the big space. The Vivarana school preferred to regard the finite individual as a reflection of Brahman in the inner sense. As the moon is one, but its reflections are many, so also Brahman is one, but its reflections are many.

Later followers of Shankara, such as Shriharsa in his *Khandanakhandakhadya* and his commentator Chitsukha, used a destructive, negative dialectic in the manner of Nagarjuna to criticize man's basic concepts about the world.

Concepts of Bhedabheda

The philosophies of transcendence and immanence (bhedabheda) assert both identity and difference between the world and finite individuals, on the one hand, and Brahman, on the other. The world and finite individuals are real and yet both different and not different from the Brahman.

Among pre-Shankara commentators on the Vedanta-sutras, Bhartrprapancha defended the thesis of bhedabheda, and Bhaskara (c. 9th century) closely followed him. Bhartrprapancha's commentary is not extant; the only known source of knowledge is Shankara's reference to him in his commentary on the Brhadaranyaka Upanisad, in which *Bhartrprapantilde*; ca is said to have held that though Brahman as cause is different from Brahman as effect, the two are identical inasmuch as the effect dissolves into the cause, as the waves return into the sea. Bhaskara viewed Brahman as both the material and the efficient cause of the world.

The doctrine of maya was totally rejected. Brahman undergoes the modifications by his own power. As waves are both different from and identical with the sea, so are the world and the finite individuals in relation to Brahman. The finite selves are parts of Brahman, as sparks of fire are parts of fire. But the finite soul exists, since beginningless time, under the influence of ignorance. It is atomic in extension and yet animates the whole body. Corresponding to the material world and the finite selves, Bhaskara ascribed to God two powers of self-modification. Bhaskara, in his theory of knowledge, distinguished between self-consciousness that is ever-present and objective knowledge that passively arises out of appropriate causal conditions but is not an activity.

Mind, thus, is a sense organ. Bhaskara subscribed to the general Vedanta thesis that knowledge is intrinsically true, though falsity is extrinsic to it. In his ethical views, Bhaskara regarded religious duties as binding at all stages of life. He upheld a theory known as Jnana-karmasamuchaya-vada: performance of duties together with knowledge of Brahman leads to liberation. In religious life, Bhaskara was an advocate of bhakti, but bhakti is not a mere feeling of love or affection for God, but rather is dhyana, or meditation, directed toward the transcendent Brahman who is not exhausted in his manifestations. Bhaskara denied the possibility of liberation in bodily existence.

The bhedabheda point of view had various other adherents: Vijnanabhiksu, Nimbarka, Vallabha, and Chaitanya.

Ramanuja (11th century) sought to synthesize a long tradition of theistic religion with the absolutistic monism of the Upanishads, a task in which he had been preceded by no less an authority than the Bhagavadgita. In his general philosophical position, he followed the vrttikara Bodhayana, the Vakyakara (to whom he referred but whose identity is not established except that he advocated a theory of real modification of Brahman), Nathamuni (c. 1000), and his own teachers' teacher Yamunacharya (c. 1050).

Ramanuja

The main religious inspirations are from the theistic tradition of the Alvar poet-saints and their commentators known as the Acharyas, who sought to combine knowledge with action (Karma) as the right means to liberation. There is also, besides the Vedic tradition, the religious tradition of Agamas, particularly of the Pancharatra literature.

It is within this old tradition that Ramanuja's philosophical and religious thought developed. Ramanuja rejected Shankara's conception of Brahman as an indeterminate, qualityless, and differenceless reality on the ground that such a reality cannot be perceived, known, thought of, or even spoken about, in which case it is nothing short of a fiction. In substantiating this contention, Ramanuja undertook, in his Shri-bhasya on the Vedanta-sutras, a detailed examination of the different ways of knowing.

Perception, either non-conceptualized or conceptualized, always apprehends its object as being something, the only difference between the two modes of perception being that the former takes place when one perceives an individual of a certain class for the first time and thus does not subsume it under the same class as some other individuals. Nor can inference provide one with knowledge of an indeterminate reality, because in inference one always knows something as coming under a general rule. The same holds true of verbal testimony.

This kind of knowledge arises from understanding sentences. For Ramanuja there is nothing like a pure consciousness without subject and without object. All consciousness is of something and belongs to someone. He also held that it is not true that consciousness cannot be the object of another consciousness. In fact, one's own past consciousness becomes the object of present consciousness. Consciousness is self-shining only when it reveals an object to its own owner—*i.e.*, the self.

Rejecting Shankara's conception of reality, Ramanuja defended the thesis that Brahman is a being with infinitely perfect excellent virtues, a being whose perfection cannot be exceeded. The world and finite individuals are real, and together they constitute the body of Brahman. The category of body and soul is central to his way of thinking. Body is that which can be controlled and moved for the purpose of the spirit. The material world and the conscious spirits, though substantive realities, are yet inseparable from Brahman and thus qualify him in the same sense in which body qualifies the soul. Brahman is spiritual-material-qualified.

Ramanuja and his followers undertook criticisms of Shankara's illusionism, particularly of his doctrine of avidya (ignorance) and the falsity of the world. For Ramanuja, such a beginningless, positive avidya could not have any locus or any object, and if it does conceal the self-shining Brahman, then there would be no way of escaping from its clutches.

A most striking feature of Ramanuja's epistemology is his uncompromising realism. Whatever is known is real, and only the real can be known. This led him to advocate the thesis that even the object of error is real—error is really incomplete knowledge—and correction of error is really completion of incomplete knowledge.

The state of Moksha is not a state in which the individuality is negated. In fact, the sense of "I" persists even after liberation, for the self is truly the object of the notion of "I." What is destroyed is egoism, the false sense of independence. The means thereto is bhakti, leading to God's grace. But by bhakti Ramanuja means dhyana, or intense meditation with love. Obligation to perform one's scriptural duties is never transcended. Liberation is a state of blessedness in the company of God. A path emphasized by Ramanuja for all persons is complete self-surrender (prapatti) to God's will and making oneself worthy of his grace. In his social outlook, Ramanuja believed that bhakti does not recognize barriers of caste and classes.

The doctrinal differences among the followers of Ramanuja is not so great as among Shankara's. Writers such as Sudarshana Suri and Venkatanatha continued to elaborate and defend the theses of the master, and much of their writing is polemical. Some differences are to be found regarding the nature of emancipation, the nature of devotion, and other ritual matters. The followers are divided into two schools: the Uttara-kalarya, led by Venkatanatha, and the Daksina-kalarya, led by Lokacharya. One of the points at issue is whether or not emancipation is destructible; another, whether there is a difference between liberation attained by mere self-knowledge and that attained by knowledge of God. There also were differences in interpreting the exact nature of self-surrender to God and the degree of passivity or activity required of the worshipper.

Madhva

Madhva (born 1199?) belonged to the tradition of Vaisnava religious faith and showed a great polemical spirit in refuting Shankara's philosophy and in converting people to his own fold. An uncompromising dualist, he traced back dualistic thought even to some of the Upanishads. His main works are his commentaries on the Upanishads, the Gita, and the Vedanta-sutras. He also wrote a commentary on the Mahabharata and several logical and polemical treatises.

He glorified difference. Five types of differences are central to Madhva's system: difference between soul and God, between soul and soul, between soul and matter, between God and matter, and that between matter and matter. Brahman is the fullness of qualities, and by his own intrinsic nature, Brahman produces the world. The individual, otherwise free, is dependent only upon God. The Advaita concepts of falsity and indescribability of the world were severely criticized and rejected. In his epistemology, Madhva admitted three ways of knowing: perception, inference, and verbal testimony. In Madhva's system the existence of God cannot be proved; it can be learned only from the scriptures.

Bondage and release both are real and devotion is the only way to release, but ultimately it is God's grace that saves. Scriptural duties, when performed without any ulterior motive, purify the mind and help one to receive God's grace. Among the other theistic schools of Vedanta, brief mention may be made of the schools of Nimbarka (c. 12th century), Vallabha (15th century), and Chaitanya (16th century).

Nimbarka

Nimbarka's philosophy is known as Bhedabheda because he emphasized both identity and difference of the world and finite souls with Brahman. His religious sect is known as the *Sanaka-Sampradaya* of Vaisnavism. Nimbarka's commentary of the Vedanta-sutras is known as *Vedanta-parijata-saurabha* and is commented on by Shrinivasa in his *Vedanta-kaustubha*. Of the three realities admitted—God, souls, and matter—God is the independent reality, self-conscious, controller of the other two, free from all defects, abode of all good qualities, and both the material and efficient cause of the world.

The souls are dependent, self-conscious, capable of enjoyment, controlled, atomic in size, many in number, and eternal but seemingly subject to birth and death because of ignorance and Karma. Matter is of three kinds: nonnatural matter, which constitutes divine body; natural matter constituted by the three gunas; and time. Both souls and matter are pervaded by God. Their relation is one of difference-with-nodifference.

Liberation is because of a knowledge that makes God's grace possible. There is no need for Vedic duties after knowledge is attained, nor is performance of such duties necessary for acquiring knowledge.

Vallabha

Vallabha's commentary on the Vedanta-sutras is known as Anubhasya ("The Brief Commentary"), which is commented upon by Purusottama in his Bhasya-prakasha ("Lights on the Commentary"). His philosophy is called pure non-dualism—"pure" meaning "undefiled by maya." His religious sect is known as the Rudra-Sampradaya of Vaisnavism and also Pustimarga, or the path of grace. Brahman, or Shri Krishna, is viewed as the only independent reality; in his essence he is existence, consciousness, and bliss, and souls and matter are his real manifestations.

Maya is but his power of self-manifestation. Vallabha admitted neither parinama (of Samkhya) nor Vivarta (of Shankara). According to him, the modifications are such that they leave Brahman unaffected. From his aspect of "existence" spring life, senses, and body. From "consciousness" spring the finite, atomic souls. From "bliss" spring the presiding deities, or antaryamins, for whom Vallabha finds place on his ontology. This threefold nature of God pervades all beings.

World is real; but samsara, the cycle of birth and death, is unreal, and time is regarded as God's power of action. Like all other Vedantins, Vallabha rejected the Vaishesika relation of samavaya and replaced it by tadatmya, or identity. The means to liberation is bhakti, which is defined as firm affection for God and also loving service (seva). Bhakti does not lead to knowledge, but knowledge is regarded as a part of bhakti. The notion of "grace" plays an important role in Vallabha's religious thought. He is also opposed to renunciation.

Chaitanya

Chaitanya (1485-1533) was one of the most influential and remarkable of the medieval saints of India. His life is characterized by almost unique emotional fervour, hovering on the pathological, which was directed toward Shri Krishna (the incarnation of Vishnu). He has not written anything, but the discourses recorded by contemporaries give an idea of his philosophical thought that was later developed by his followers, particularly by Rupa Gosvamin and Jiva Gosvamin. Rupa is the author of two great works: *Bhakti-rasamrta-sindhu* ("The Ocean of the Nectar of the Essence of Bhakti") and *Ujjvalanilamani* ("The Shining Blue Jewel"). Jiva's main work is the great and voluminous Satsamdarbha.

These are the main sources of the philosophy of Bengal Vaisnavism. Chaitanya rejected the conception of an intermediate Brahman. Brahman, according to him, has three powers: the transcendent power that is threefold (the power of bliss, the power of being, and the power of consciousness) and the two immanent powers, namely, the powers of creating souls and the material world. Jiva Gosvamin regarded bliss to be the very substance of Brahman who, with the totality of all his powers, is called God. Jiva distinguished between God's essential power, his peripheral power that creates the souls, and the external power (called maya) that creates cosmic forms. The relation between God and his powers is neither identity nor difference, nor identity-with-difference.

This relation, unthinkable and suprarational, is central to Chaitanya's philosophy. For Jiva, the relation between any whole and its parts is unthinkable. Bhakti is the means to emancipation. Bhakti is conceived as a reciprocal relation between man and God, a manifestation of God's power in man. The works of Jiva and Rupa delineated a detailed and fairly exhaustive classification of the types and gradations of bhakti.

Vedanta in Daily Life

The Upanishads form the basis of Vedanta. Vedanta is the basic culture of India. It is the national philosophy of India.

It is the Moksha Sastra or the Science of Emancipation. Absolutism is the pivot of Vedanta philosophy.

Vedanta upholds the reality of the indivisible, immanent and transcendent Spirit. It does not exclude matter. It does not exclude anything. The oneness of all existence is the message which Vedanta teaches. It has kept Hindu society alive for the past several thousand years.

Vedanta is the only bold philosophy which dares call man God, not merely the son of God, or His servant. It proclaims with emphasis that you are the immortal, all-pervading *Atman*, the Universal Soul or Supreme Brahman in essence, in reality. Boldness is the key-note of Vedanta. The message of Vedanta is fearlessness, soul-force and unity of consciousness.

Vedanta does not ask for converts or proselytes, but a deeper reassessment of the divine-human equation, a return to the fundamental question of every being: "What am I really ? What is my real Self ?" Vedanta proclaims: "Man, in essence, is identical with the Supreme Being."

Vedanta denotes one's identity with the rest of humanity. According to Vedanta, there is no stranger in this world. Everyone is related to one another in the kinship of the Spirit. In Vedanta, there is no 'mine' and 'for me'; but 'ours' and 'for us'; and ultimately, 'His' and 'for Him'. If the Vedanta philosophy is rightly understood and acted up to, then it will obliterate all evils that emanate from factional and racial prejudices. Vedanta is no creed, no ceremony or form of worship. It is the science of right living. It is not the sole monopoly of the Hindus or the recluses. It is for all.

Vedanta has no quarrel with any religion whatsoever. It preaches universal principles. Vedanta is the only universal, eternal religion. It is a great leveler. It unites all. It gives room to all.

Vedanta encloses within its sphere all the religions of the world and is strong enough to make them all useful and enduring. Vedanta never interferes with forms. It concerns itself solely with the life of religions. The Christian need not renounce his Christianity, the Buddhist may stick to his Noble

Eightfold path, the Muslim may stick to his Quran, and yet all these may follow the Vedanta and realize in practice all its high ideals and truths. Their love to their respective prophets and Bibles will become more sober, more enlightened, and more enduring. Religious animosity will vanish and the world will move on to its great end without any friction, with greater dignity and more goodwill among its denizens.

Vedanta means no slavery. It gives freedom to all. It never condemns any man as beyond hope, never looks upon anyone as accused, but takes all mankind within its fold. Vedanta is extremely catholic and liberal in its outlook. Vedanta can offer to the modern society a common faith, a common body of principles, and a common moral discipline. It is highly scientific in outlook and has a real appeal to men and women of today.

There is no philosophy so bold and sublime as the philosophy of the Vedanta. It is Vedanta alone that can eradicate totally human sufferings and can bring everlasting peace and happiness. Even a little understanding and a little practice of Vedanta can raise a man to magnanimous heights of Brahmanhood or God-consciousness and remove all sorts of fears, worries and anxieties of this mundane life.

Some ignorant people only say that Vedanta preaches immorality, hatred and pessimism. This is a very sad mistake. Vedanta does not preach either immorality or even indifference to morality. Vedanta wants you to destroy Moha or selfish love and passion for the body, and develop pure, disinterested cosmic love or the magnanimous divine Love. Vedanta never preaches pessimism, but it preaches the pinnacle of optimism. Licentiousness is mistaken for a life of expansion. If a man can eat anything in any hotel in any part of the world, if he can move socially with any man or woman, that does not mean he is a Vedantin. There is much tall talk of Vedanta nowadays. People talk of unity, oneness and equality, but fight out for little, useless things. They are full of jealousy and hatred. I cannot imagine this. I am simply stunned.

I believe in practical Vedanta. I believe in solid spiritual practices. I believe in thorough overhauling of worldly nature, worldliness of various sorts.

You must be a practical Vedantin. You should live in the spirit of Vedanta. Mere theorizing and lecturing is only intellectual gymnastics and lingual warfare. This will not suffice. What is the use of reading too many books on Vedanta like Chit-Sukhi, Khandana Khanda Khadyam, etc. ? You must radiate love to one and all. The spirit of Vedanta must be ingrained in your cells, tissues, veins, nerves and bones. It must become part and parcel of your nature. You must think of unity, speak of unity and act in unity.

The sun, the Ganga, the flowers, the sandal trees, the fruit-bearing trees, the cows-all teach practical Vedanta to the world. They live for serving the humanity in a disinterested spirit. The sun radiates its light alike over a cottage of a peasant and a palace of a king. The cool refreshing waters of the Ganga are drunk by all. The flowers waft their fragrance to all without expecting anything. The sandal tree wafts its aroma even to the man who cuts it with an axe. All fruit-bearing trees behave in the same manner. O selfish, ignorant man ! Learn lessons from these practical Vedantins and become wise.

Vedanta does not preach a doctrine of negation of human effort. It wants you to have a changed mental attitude. It demands a changed angle of vision. Till now, the world was everything. Hereafter, the Reality alone is everything.

Once there lived two friends, Ram and Gopal. They were both philosophers. By analysis and self-enquiry, Ram learned to see the Glory of the Supreme Self reflected in and through all the universe. But Gopal continued to remain a theoretical philosopher, condemning the universe as an illusion and dream containing nothing but evil and vice.

One day, after a long time, Ram called on his friend. Gopal discussed, for a long time, as usual, the evil in this universe, and in the end asked Ram what present he had brought for his friend. Ram, after thinking a while, produced a broken piece of mirror from his pocket and handing it over to Gopal, said, "This is my little and humble present. It will help you to understand your own beauty and charm, which you cannot otherwise see."

Gopal learnt a lesson, and from that moment began to visualize and understand the Glory of the Supreme Self reflected in all the universe. Nothing is useless in this world. The non-self exists to reflect and glorify the Self. Otherwise how can you know the existence of the Self ? Verily, the non-self is the mirror that truly reflects the Self for us to cognize.

So, too, evil is the mirror for good. The presence of sages and saints is easily cognized amidst an assembly of ignorant men. Learn to see the good reflected by the evil, and say, "Evil exists to remind me of good, the perishable exists to remind me of the Imperishable," and so on.

Truly, this universe is a mirror that reminds us of God. Learn not to condemn it as an illusion and dream, but to utilize it to feel the presence of God.

Learn to discriminate between the permanent and the impermanent. Behold the Self in all beings, in all objects. Share what you have, physical, mental, moral or spiritual, with all. Serve the Self in all. Feel when you serve others, that you are serving your own Self. Love thy neighbour as thyself. Melt all illusory differences. Remove all barriers that separate man from man. Mix with all. Embrace all. Destroy the sex-idea and body-idea by constantly thinking on the Self or the sexless, bodiless *Atman*. Fix the mind on the Self when you work. This is practical Vedanta. This is the essence of the teachings of the Upanishads and sages of yore. This is real, eternal life in *Atman*. Put these things in practice in the daily battle of life. You will shine as dynamic Yogin or Jivanmukta. There is no doubt of this.

The Absolute Reality which is the fundamental cause of the universe and all beings is full (beyond all limitations which can be caused by time, space, and material). It is pure, free indivisible and homogeneous Consciousness. The individual self *Jiva)* which appears limited by time, space, and material, or name and form, is also full, pure, free, indivisible and homogeneous Consciousness. The individual Consciousness (self) which is full has come from that Absolute Consciousness which is also full. Here comes a discrepancy. How can something full come from something that is also full and still both remain full? This disparity is solved by the illustration of space. Space is recognized as full, free, pure, boundless, unattached, all-pervading, indivisible and homogeneous. When a barrier is erected in space, we say the space is divided and call it "room space, limited space,", etc. If there is dust or contamination in the room, we say the space in the room is dusty or impure. We all know that space cannot be divided or become impure.

It is always full, free, pure and indivisible. The divisions are assumed due to the barriers we construct in space. Similar is the nature of the Absolute Consciousness which is the Supreme Reality. Nothing can ever come from it. It alone exists. The separate individuality which one feels is due to the limiting adjuncts of body-mind complex which are, in reality, illusory appearances like dream objects or mirage seen in a desert or a rope mistakenly seen as a snake in dim light. When viewed in its real nature as Pure Consciousness alone, divested of its limitations, the individual Self is known to be one and the same Absolute Consciousness.

Vedanta is the teachings of the Upanishads (the concluding chapters of the four Vedas), the Brahma Sutras, the Bhagavad Gita, and their commentaries by Shri Shankaracharya, the 8th century philosopher saint. *Veda* means "knowledge" and *anta* means "end". Thus Vedanta means "the end of knowledge".

Sub-schools of Vedanta

Advaita Vedanta: Advaita Vedanta is the most influential school of all, and many philosopers, both Indian and Western, have been influenced by it. It was propounded by Adi Shankara, a great Hindu philosopher, and his Paramaguru Gaudapada, who described Ajativada. According to this school of Vedanta, Brahman is the only reality, and the world, as it appears, is illusory. As Brahman is the sole reality, it cannot be said to possess any attributes whatsoever. An illusionary power of Brahman called My causes the world to arise. Ignorance of this reality is the cause of all suffering in the world and only upon true knowledge of Brahman can liberation be attained. When a person tries to know Brahman through his mind, due to the influence of My, Brahman appears as God (Ishvara), separate from the world and from the individual. In reality, there is no difference between the individual soul *j+vtman* (see *Atman*) and Brahman. Liberation lies in knowing the reality of this non-difference (*i.e.* a-dvaita, "not-two" ness). Thus, the path to liberation is finally only through knowledge (*jnna*).

Vishishtadvaita: Vishishtadvaita was propounded by Ramanuja and says that the *j+vtman* is a part of Brahman, and hence is similar, but not identical. The main difference from Advaita is that in Visishtadvaita, the Brahman is asserted to have attributes, including the individual conscious souls and matter. Brahman, matter and the individual souls are distinct but mutually inseparable entities. This school propounds Bhakti or devotion to God visualized as Vishnu to be the path to liberation. My is seen as the creative power of God.

Dvaita: Dvaita was propounded by Madhva. It identifies God with Brahman completely, and in turn with Vishnu or his incarnation Krishna. It regards Brahman, all individual souls (*j+vtmans*) and matter as eternal and mutually separate entities. This school also advocated Bhakti as the route to liberation. There is no concept of My as an illusionary power behind the world.

Dvaitdvaita: Dvaitdvaita was propounded by Nimbrka, based upon an earlier school called Bhedbheda, which was taught by Bhaskara. According to this school, the *jivtman* is at once the same as yet different from Brahman.Jiva relation may be regarded as dvaita from one point of view and Advaita from another.

ShuddhAdvaita: ShuddhAdvaita propounded by Vallabha. This system also encouraged Bhakti as the only means of liberation to go to Goloka (lit., the world of cows). The world is said to be the sport (Leela) of Krishna, who is *Sat-Chit-Ananda*.

Achintya Bhedbheda: Achintya Bhedbheda propounded by Chaitanya Mahaprabhu. This doctrine is followed by the world famous ISKCON movement.

While Adi Shankara propounded the Smrta denomination, all the other above-mentioned acharyas were strongly

Vaishnavite in orientation. The epistemology of Advaita, Vishishtadvaita and Mimamsa (*i.e.*, Purva) is common.

Purnadvaita or Integral Advaita: Sri Aurobindo, in his The Life Divine, has synthesized all the exant schools of Vedanta and given the most comprehensive resolution integrating cues from the Western metaphysics and modern science. He has also restored the umbilical cord of the Vedantic exegesis with the Vedas. For instance, his strong difference with Adi Shankara over the interpretation of *Ishavasyam* is interesting.

The Advaita Parampara

Narayanam padmabhuvam vasishtham. Shaktim. ca tatputra parasaram. ca
vyasam. Sukam. Gaudapadam. mahantam. govindayogindram athasya Sishyam.
Sri Shankaracharyan athasya Padmapadam. ca Hastamalakam. ca Sishyam.
tam. totakam. varttikakaramanyan asmad gurun santatamanatosmi
Sadasiva samarambham. Shankaracharya madhyamam.
asmadacharya paryantam. vande guru paramparam.

These are the Advaita *guru parampara* verses, which salute the prominent gurus of Advaita, starting from narayana through Shankara and his disciples, upto the Acharyas of today. It is typical of Advaita that the first guru is called *narayana* (Vishnu) in the first verse and Sadasiva (Shiva) in the second. The parampara thus lists:

Narayana
Padmabhuva (brahma)
Vasishtha
Shakti
Parasara
Vyasa
Suka
Gaudapada

Govinda Bhagavatpada

Shankaracharya

Padmapada, Hastamalaka, Totaka, Suresvara (Varttikakara),

and others (anya:).

In the Indian religious and philosophical traditions, all knowledge is traced back to the Gods and to the Rishis who saw the vedas.

Thus, the Advaita *Guru-parampara* begins with the *Daiva-parampara*, followed by the *Rishi-parampara*, which includes the vedic seers vasishtha, Shakti, parasara, his son Vyasa, (the famous redactor of the vedas, he is also traditionally identified with Badarayana, the composer of the Brahmasutras), and vyasa's son Suka. After Suka, we turn to the *Manava-parampara*, which brings us to historical times and personalities.

The traditions regarding these human gurus are recorded in the *Shankaravijaya* literature, and typically, they are regarded as incarnations of various deities Gaudapada is the famous author of the *Mandukya Karikas* that are attached to the *Mandukya Upanishad*.

His disciple, Govinda, is regarded as an incarnation of Adisesha, the cosmic serpent. He was the preceptor of Shankara, who is regarded as an incarnation of Shiva. Shankara's four well-known disciples were named *Padmapada, Hastamalaka*, totaka and Suresvara (Varttikakara). Tradition has it that Shankara appointed these four disciples as heads of the four Mathas that he founded. The others are the gurus who come later in the tradition.

The doctrine of Advaita Vedanta as expounded by Shankara can be summed up in half a verse: *"Brahma Satyam Jagan Mithya Jivo Brahmaiva Na Aparah"*—Brahman (the Absolute) is alone real; this world is unreal; and the Jiva or the individual soul is non-different from Brahman. This is the quintessence of his philosophy.

According to Sri Shankara, whatever is, is Brahman. Brahman Itself is absolutely homogeneous. All difference and plurality are illusory.

Tenets of Advaita Vedanta

- Brahman (the Absolute) is alone real; this world is unreal; and the Jiva or the individual soul is non-different from Brahman.
- The *Atman* is self-evident (Svatah-siddha). It is not established by extraneous proofs. It is not possible to deny the *Atman*, because it is the very essence of the one who denies It.
- Brahman is not an object, as it is Adrisya, beyond the reach of senses, mind or intellect. It is not another. It is all-full, infinite, changeless, self-existent, self-delight, self-knowledge and self-bliss. It is Svarupa, essence. It is the essence of the knower. It is the Seer (Drashta), Transcendent (Turiya) and Silent Witness (Sakshi). It is always the Witnessing Subject. It can never become an object as it is beyond the reach of the senses. Brahman is non-dual, one without a second. It has no other beside it.
- Sat-Chit-Ananda constitute the very essence or Svarupa of Brahman, and not just its attributes.
- The world is not an illusion according to Shankara. The world is relatively real (Vyavaharika Satta), while Brahman is absolutely real (Paramarthika Satta). The unchanging Brahman appears as the changing world because of a superimposition of non-Self (objects) or Self (subject-Brahman). This is called Avidya.
- The Jiva or the individual soul is only relatively real. Its individuality lasts only so long as it is subject to unreal Upadhis or limiting conditions due to Avidya. The Jiva identifies itself with the body, mind and the senses, when it is deluded by Avidya or ignorance. Just as the bubble becomes one with the ocean when it bursts, so also the Jiva or the empirical self becomes one with Brahman when it gets knowledge of Brahman. When knowledge dawns in it through annihilation of Avidya, it is freed from its individuality and finitude and realizes its essential Sacchidananda nature. It

merges itself in the ocean of bliss. The river of life joins the ocean of existence. This is the Truth.

- Because Samsara (or duality) exists due to ignorance or Avidya, Knowledge (Jnana) alone can make an individual realize his true nature. Karma Yoga, Bhakti Yoga, Raja Yoga, etc., are necessary only to purify the individual and to help remove this Avidya. All other paths culminate in Jnana.
- Brahma Jnana is not about acquiring any external knowledge (as Brahman can't be an object of knowledge), it just about removing the Avidya or Maya.

Brahma Sutras (Vedanta Sutras)

ADHYAYA 1

Adhikarana/ Sutra

PADA 1

1. Then, therefore, enquiry into Brahman.
2. From which springs the origination.
3. Scripture alone is the means.
4. All the scriptures' ultimate goal is Brahman.

5-11. Non intelligent pradhana (of Sankhya) cannot be the (seeing and willing) cause of the world.

6-12. 'Anandamaya' mentioned in the (Taittiriya) Upanisad is the Supreme Lord, not Jiva.

20-21. The golden person seen within the sun and within the eye, mentioned in the Chandogya Upanisad (1.6) is not a Jiva of high eminence, but the Supreme Lord.

22. The ether mentioned in Ch. 1.6 is not the elemental ether but the Supreme Lord.

23. The prana mentioned in Ch. 1.11.5 denotes the Supreme Lord.

24-27. The light spoken of in Ch. 3.13.7 is not physical light, but the Supreme Lord.

28-31. The prana mentioned in *Kausitaki* 3.2. is the Supreme Lord.

PADA 2

1-8. 'Manomaya' and *'prana sarira'* mentioned in Ch 3.14.1,2 is not Jiva, but the Supreme Lord.

9-10. The Being to whom Brahmanas and Kshatriyas are food, referred to in Katha 1.2.24 is the Supreme Lord.

11-12. The two entered into the cave-Katha 1.3.1-are the Supreme Lord and Jiva.

13-17. The person within the eye-Ch. 4.15.1 is the Supreme Lord.

18-20. *'Antaryamin'* mentioned in Br. 3.7.3 is the Supreme Lord (not *pradhana* nor *Jiva*).

21-23. Being described as possessed of invisibility in Mund. 1.1.56 is the Supreme Lord (not pradhana or Jiva).

24-32. *'Atma vaisvanarah'* in Ch. 5.11.6 is the Supreme Lord.

PADA 3

1-7. Being within whom heaven and earth are woven, as in Mund. 2.2.5 is the Supreme Lord, because of the term 'atman' (not *pradhana* or *Jiva*).

8-9. 'Bhuman' (the great) in Ch. 7.23.24 is the Supreme Lord (not prana).

10-12. 'Aksara' (imperishable) in Br. 3.88 is the Supreme Lord.

13. *'Para purusa'* to be meditated upon by 'om' in Prasna. 5.5 is the Supreme Lord (because it is described as object of realization).

14-23. 'Dahara akasah' within the lotus of the heart in Ch. 8.1.1. is not elemental ether or Jiva but the Supreme Lord (-grammatical arguments!-).(22 & 23 as 6) by Shankara.

24-25. The thumb size Being in Katha 2.4.13 is not Jiva, but the Supreme Lord.

26-33. Gods qualified for Brahmavidya? (animals already denied in 7)

34-38. Sudras denied qualification.

39. The prana before whom everything trembles as in Katha 2.6.2 is the Supreme Lord.

40. Jyoti mentioned in Ch. 8.12.3 is the Supreme Lord.

41. The ether that reveals names and forms as in Ch. 8.14.1 is neither the elemental ether, nor the Jiva, but the Supreme Lord.

42-43. *Vijnana maya* in Br. 4.3-7 is not Jiva but the Supreme Lord.

PADA 4

1-7. *Avyakta* in Katha 1.3.10,11 does not denote pradhana, but the subtle body (and gross body?).

8-10. *Tri rupa aja* in Svet. 4.5 cannot be Pradhana or Prakrti but the transcendental energy of the Supreme Lord who is the cause of all causes.

11-13. *Pancha pancha janah* in Br. 4.4.17 are not the twenty five principles but the five pranas.

14-15. The Supreme intelligent Lord is the cause of the world not the individual soul, but the Supreme Lord.

23-27. The Supreme Lord is not only the nimitta karana (efficient cause) but upadana (material cause) as well.

28. All the passages in Vedanta are referring to the Supreme Lord only.

ADHYAYA 2

PADA 1

1-2. Sankhya rejected because it is against Shruti.

3. Yoga rejected.

4-11. Objection that non intelligent world cannot be caused by intelligent Supreme Lord (Brahman) is answered. By the effect produced the cause, Lord, is not contaminated.

12. Atomic theory refuted like above.

13. The enjoyer Lord and the enjoyed Jiva are qualitatively one.

14-20. Vaisesika doctrine of effect as something new created from the cause is refuted. The effect is only transformation of the energies of the Lord.

21-23. Brhad 2.4.5, Ch.8.7.1, Br.4.3.35-Jiva is simultaneously one and different from the Lord

24-25. The Lord creates the world by His mere will.

26-29. The Lord creates the world without entirely passing over it, and at the same time remaining one and undivided.

30-31. The Lord, unassisted, can create the world through His variegated transcendental potencies.

32-33. He has no self interest in creating the world (lila).

34-36. The Lord is not to be blamed for the happiness and distress of the jivas—He sanctions results according to the *merits-samsara* is beginningless.

37. The Lord is omniscient and omnipotent and capable of creating the world.

PADA 2

1-10. Refutation of *Sankhya*; non intelligent *pradhana* cannot possibly create or proceed with any activity without the guidance of an intelligent being.

11-17. Refutation of *Vaisesika*; atoms as the cause of creation defeated.

18-32. Refutation of Bauddhas (Buddhists) *bahyarthavadins* (both the material objects *sarvastitvavadins* and their cognitions are real) (realists) *vijnanavadins* (ideas only are real *Yogacharas* external objects apart from (idealists) their cognitions have no independent existence) *sunyavadis* (everything is void) (nihilists)

33-36. Refutation of Jaina.

37-41. Refutation of Pasupata (refuting that Ishvara is only the efficient cause).

42-45. Refutation of Sakta.

PADA 3

1-7. Earth is not co-eternal with Lord but the first effect.

8. Air from earth.

9. The Lord has no cause.

10-12. Fire from air, water from fire, earth from water.

13. The Lord is the cause of these elements but He is not affected by these elements.

14. The reabsorbtion of the elements is inverse to this order.

15. This is not disturbed by the organs and senses because they are created and destroyed with the elements which they consist.

16. The birth and death belong to the body and not to the Jiva.

17. Jiva is never created.

18. Knowledge is a constitutional character of Jiva.

19-32. Jiva is infinitesimal, has knowledge as his attribute.

33-40. Kartrtva is a natural attribute of Jiva.

41-42. This kartrtva of the Jiva depends upon the Lord.

43-53. The relationship between Jiva and the Lord

PADA 4

1-4. Pranas are created from the Lord.

5-6. They are eleven in number (Jnana, Karma-indriyas and intelligence).

7. They are anu (subtle).

8. Attributes of pranas transferred (atidesa) to chief vital air.

9-12. Nature and character of the chief vital air.

13. It is subtle (anu).

14-16. It is controlled by the Supreme Lord.

17-19. Pranas are not modifications of the chief vital air, they are independent principles.

20-22. Activity of creation belongs to the Super soul not to the Jiva.

ADHYAYA 3

PADA 1

1-7. Individual Jiva in obtaining a different body continues with the same subtle body.

8-11. The souls who have enjoyed results of pious activities in the moon descend to earth-with an amsa of their previous pious *karmas* for which results cannot be enjoyed in the moon, but must be enjoyed in this planet only—which determines the design of the new body.

12-21. Those who have knowledge (*Vidya*) and who perform sacrifices ascend to moon. For the rest there is the third place, Yama's abode.

22. The subtle bodies descending from moon, through the ether, air, etc., do not become identical with them, but only resemble them.

23. The entire descent of the soul occupies a very short time only.

24-27. When the souls finally enter into plants and so on they do not participate in the life of the plants but are merely in external contact with them.

PADA 2

1-6. The creation of the dreaming state is also willed by the Supreme Lord—they are real to the extent that they are indications of future (or recollections of past).

7-8. In dreamless sleep the Jiva abides with Super soul within the heart.

9. The soul awakening from the sleep is the same as before, otherwise he won't be remembering past experiences and injunctions of sacrifices.

10. Swoon is midway between sleep and death.

11-21. Supreme Lord is *nirvisesa* because He is *prakrta rupa hina*. The attributes of Jiva and matter do not apply to the Lord. He is not materially affected, being transcendental.

22-26. Supreme Lord possesses infinite number of transcendental attributes beyond limitations.

27-30. Matter is transformation of the Lord's energy-so it is one and in the same time different from Him.

PADA 3

1-4. The cognitions of the Lord, taught by all the Vedantic texts, are identical on account of the non-difference of injunctions (*upasita-vidyat*, etc.).

5. Since the goal is the Supreme Lord wherever there are contradictions and different kinds of processes mentioned in the scriptures, *upasamhara* (tallying) is necessary-actually the process is one and the same. Sandilya Vidya in three passages is one and the same (though various details differ). *Vaisvanara Vidya* (Ch. 5.12.2, *Satapatha* 10.6.1-11) is identical-the process, the goal are same. *Katha* 1.2.15-that word which all the Vedas record. Aitareya 3.2.3.12-Him only the bahvrcas consider in the great hymn, the Adhvaryus in the sacrificial fire, the Chandogas in the mahavrata ceremony.

6-8. The subject of chandogya Vidya (1.1.3) is only the sacred om while the Brhad 1.3 represents the whole udgitha as goal.

9-10. Are all qualities to be included in meditation on the Lord?

11-13. Only the essential and unalterable attributes such as ananda and Jnana are to be taken into account everywhere.

14-26. similar subject.

27-31. Regarding the soul's freeing himself on death and on liberation.

32-54. Regarding the synthesis of different processes considering the goal.

PADA 4

1-50. The knowledge of Brahman is not subordinate to action, but independent. The ashrama karmani are obligatory for him who does not strive for *mukti* *balya*, *panditya* and *mauna* are three conditions enjoined for the *Sannyasin*.

51-52. Liberation is either immediately after death or gradual according to the *sadhana*.

ADHYAYA 4

PADA 1

1-2. Meditation on the Lord as enjoined in the scriptures is to be repeated again and again until the realization of the final goal takes place.

3. The goal of meditation is the Lord who is the soul of all.

4. Regarding *pratikopasana*.

5. Supreme Lord is the only object of meditation.

6. In angopasanas the Lord who is ultimately the controller of everything should be the goal.

7-10. Regarding time and place and period of meditation.

11. It should be done favorably beyond conditions.

12. Till death the process should be continued.

13. On liberation no more control by Karma.

14. -same (pious Karma)-

15. Devotional activities prevail

16-17. They promote the origination of knowledge.

19. The realized vidvan may have to pass through several embodied existences before the *arabdhakarya karmas* will have their effects completely exhausted.

PADA 2

1-7. The path of liberation after death.

8-11. Till total liberation the subtle body continues.

12-14. Similar.

15-16. On liberation the elements merge with elements.

17. The conditioned Jiva passes from the heart out of the body. The Jiva liberated passes through the head, through susumna, the 101st vein.

18-19. The liberated soul even if departs at night passes through a ray of light through the sun.

20-21. Even if departing in daksinayana the liberated soul attains the supreme abode.

PADA 3

1-3. Reconciliation of different accounts given in the *Upanishads* (Ch. 5.10.1, Kaus. 1.3, Br. 6.2.15, etc.) as to the stations of this path and establish the path (deva-yana as opposed to pitr-yana). The stages are Agni, the day, the bright half of the month and *uttarayana*, the year, *Vayu*, *Aditya*, *chandramas*, lightning, *Varuna*, *Indra*, *Prajapati*, and the *Brahman*.

4-6. These places mean the controlling deities.

7-16. Considering the different opinions of the philosophers regarding who is led to what goal.

PADA 4

1-3. The soul on liberation attains to its original nature and does not acquire new attributes.

4. The liberated soul realizes itself as qualitatively one with the Lord.

5-7. Nature of the released soul-has transcendental attributes which are constitutional.

8-9. All his desires are satisfied by exercise of his free will properly.

10-14. The released soul attains a spiritual body (attains body of its own will).

15-16. The liberated soul on attaining constitutional position can exist in many transcendental bodies.

17-22. The liberated soul participates in all the opulences of the Lord, except the power of creating and sustaining the world. He goes back home back to Godhead and never returns.

Two worlds there are in which we live, one the external, the other internal. Human progress has been made, from days of yore, almost in parallel lines along both these worlds. The search began in the external, and man at first wanted to get answers for all the deep problems from outside nature. Man wanted to satisfy his thirst for the beautiful and the sublime from all that surrounded him; he wanted to express himself and all that was within him in the language of the concrete; and grand indeed were the answers he got, most marvellous ideas of God and worship, and most rapturous expressions of the beautiful. Sublime ideas came from the external world indeed.

But the other, opening out for humanity later, laid out before him a universe yet sublimer, yet more beautiful, and infinitely more expansive. In the Karma Kanda portion of the Vedas, we find the most wonderful ideas of religion inculcated, we find the most wonderful ideas about an overruling Creator, Preserver, and Destroyer of the universe presented before us in language sometimes the most soul-stirring. Most of you perhaps remember the most wonderful Shloka in the *Rig-Veda Samhita* where you get the description of chaos, perhaps the sublimest that has ever been attempted yet.

In spite of all this, we find it is only a painting of the sublime outside, we find that yet it is gross, that something of matter yet clings to it. Yet we find that it is only the expression of the Infinite in the language of matter, in the language of the finite, it is the infinite of the muscles and not of the mind; it is the infinite of space and not of thought. Therefore in the second portion of *Jnana Kanda*, we find there is altogether a different procedure. The first was a search in external nature for the truths of the universe; it was an attempt to get the solution of the deep problems of life from the material world. "Whose glory these Himalayas declare". This is a grand idea, but yet it was not grand enough for India.

The Indian mind had to fall back, and the research took a different direction altogether; from the external the search came to the internal, from matter to mind. There arose the cry, "When a man dies, what becomes of him?" "Some say that he exists, others that he is gone; say, O king of Death, what is the truth?" An entirely different procedure we find here. The Indian mind got all that could be had from the external world, but it did not feel satisfied with that; it wanted to search further, to dive into its own soul, and the final answer came.

The Upanishads, or the Vedanta, or the Aranyakas, or Rahasya is the name of this portion of the Vedas. Here we find at once that religion has got rid of all external formalities. Here we find at once that spiritual things are told not in the language of matter, but in the language of the spirit; the superfine in the language of the superfine. No more any grossness attaches to it, no more is there any compromise with things of worldly concern. Bold, brave, beyond the conception of the present day, stand the giant minds of the sages of the Upanishads, declaring the noblest truths that have ever been preached to humanity, without any compromise, without any fear.

This, my countrymen, I want to lay before you. Even the *Jnana Kanda* of the Vedas is a vast ocean; many lives are necessary to understand even a little of it. Truly has it been said of the Upanishads by Ramanuja that they form the head, the shoulders, the crest of the Vedas, and surely enough the Upanishads have become the Bible of modern India. The Hindus have the greatest respect for the *Karma Kanda* of the Vedas, but, for all practical purposes, we know that for ages by Shruti has been meant the Upanishads, and the Upanishads alone.

We know that all our great philosophers, whether *Vyasa*, *Patanjali*, or *Gautama*, and even the father of all philosophy, the great *Kapila* himself, whenever they wanted an authority for what they wrote, every one of them found it in the Upanishads, and nowhere else, for therein are the truths that remain for ever.

There are truths that are true only in a certain line, in a certain direction, under certain circumstances, and for certain times—those that are founded on the institutions of the times. There are other truths which are based on the nature of man himself, and which must endure so long as man himself endures. These are the truths that alone can be universal, and in spite of all the changes that have come to India, as to our social surroundings, our methods of dress, our manner of eating, our modes of worship—these universal truths of the Shrutis, the marvellous Vedantic ideas, stand out in their own sublimity, immovable, unvanquishable, deathless, and immortal.

Yet the germs of all the ideas that were developed in the Upanishads had been taught already in the Karma Kanda. The idea of the cosmos which all sects of Vedantists had to take for granted, the psychology which has formed the common basis of all the Indian schools of thought, had there been worked out already and presented before the world. A few words, therefore, about the Karma Kanda are necessary before we begin the spiritual portion, the Vedanta; and first of all we like to explain the sense in which we have used the word Vedanta.

Unfortunately there is the mistaken notion in modern India that the word Vedanta has reference only to the Advaita system; but you must always remember that in modern India the three Prasthanas are considered equally important in the study of all the systems of religion. First of all there are the Revelations, the *Shrutis*, by which we mean the Upanishads. Secondly, among our philosophies, the Sutras of Vyasa have the greatest prominence on account of their being the consummation of all the preceding systems of philosophy.

These systems are not contradictory to one another, but one is based on another, and there is a gradual unfolding of the theme which culminates in the Sutras of Vyasa. Then, between the Upanishads and the Sutras, which are the systematising of the marvellous truths of the Vedanta, comes in the Gita, the divine commentary of the Vedanta.

The Upanishads, the Vyasa-*Sutras*, and the *Gita*, therefore, have been taken up by every sect in India that wants to claim authority for orthodoxy, whether dualist, or *Vishishtadvaitist*, or *Advaitist*; the authorities of each of these are the three Prasthanas. We find that a Shankaracharya, or a Ramanuja, or a Madhvacharya, or a Vallabhacharya, or a Chaitanya—any one who wanted to propound a new sect—had to take up these three systems and write only a new commentary on them.

Therefore it would be wrong to confine the word Vedanta only to one system which has arisen out of the Upanishads. All these are covered by the word Vedanta. The Vishishtadvaitist has as much right to be called a Vedantist as the Advaitist; in fact we will go a little further and say that what we really mean by the word Hindu is really the same as Vedantist. We want you to note that these three systems have been current in India almost from time immemorial; for you must not believe that Shankara was the inventor of the Advaita system.

It existed ages before Shankara was born; he was one of its last representatives. So with the Vishishtadvaita system; it had existed ages before Ramanuja appeared, as we already know from the commentaries he has written; so with the dualistic systems that have existed side by side with the others. And with my little knowledge, we have come to the conclusion that they do not contradict each other.

Just as in the case of the six Darshanas, we find they are a gradual unfolding of the grand principles whose music beginning far back in the soft low notes, ends in the triumphant blast of the Advaita, so also in these three systems we find the gradual working up of the human mind towards higher and higher ideals till everything is merged in that wonderful unity which is reached in the Advaita system. Therefore these three are not contradictory. On the other hand we are bound to tell you that this has been a mistake committed by not a few. We find that an Advaitist teacher keeps intact those texts which especially teach Advaitism, and tries to interpret the dualistic or qualified non-dualistic texts into his own meaning.

Similarly we find dualistic teachers trying to read their dualistic meaning into Advaitic texts. Our Gurus were great men, yet there is a saying, "Even the faults of a Guru must be told". I am of opinion that in this only they were mistaken. We need not go into text-torturing, we need not go into any sort of religious dishonesty, we need not go into any sort of grammatical twaddle, we need not go about trying to put our own ideas into texts which were never meant for them, but the work is plain and becomes easier, once you understand the marvellous doctrine of *Adhikarabheda*.

It is true that the Upanishads have this one theme before them: "What is that knowing which we know everything else?"

In modern language, the theme of the Upanishads is to find an ultimate unity of things. Knowledge is nothing but finding unity in the midst of diversity. Every science is based upon this; all human knowledge is based upon the finding of unity in the midst of diversity; and if it is the task of small fragments of human knowledge, which we call our sciences, to find unity in the midst of a few different phenomena, the task becomes stupendous when the theme before us is to find unity in the midst of this marvellously diversified universe, where prevail unnumbered differences in name and form, in matter and spirit—each thought differing from every other thought, each form differing from every other form.

Yet, to harmonise these many planes and unending *Lokas*, in the midst of this infinite variety to find unity, is the theme of the Upanishads. On the other hand, the old idea of *Arundhati Nyaya* applies. To show a man the fine star Arundhati, one takes the big and brilliant nearest to it, upon which he is asked to fix his eyes first, and then it becomes quite easy to direct his sight to Arundhati. This is the task before us, and to prove my idea I have simply to show you the Upanishads, and you will see it. Nearly every chapter begins with dualistic teaching, Upasana.

God is first taught as some one who is the Creator of this universe, its Preserver, and unto whom everything goes at last. He is one to be worshipped, the Ruler, the Guide of nature, external and internal, yet appearing as if He were outside of nature and external. One step further, and we find the same teacher teaching that this God is not outside of nature, but immanent in nature. And at last both ideas are discarded, and whatever is real is He; there is no difference. "Svetaketu, That thou art." That Immanent One is at last declared to be the same that is in the human soul. Here is no compromise; here is no fear of others' opinions. Truth, bold truth, has been taught in bold language, and we need not fear to preach the truth in the same bold language today, and, by the grace of God, we hope at least to be one who dares to be that bold preacher.

To go back to our preliminaries. There are first two things to be understood—one, the psychological aspect common to all the Vedantic schools, and the other, the cosmological aspect. We will first take up the latter. Today we find wonderful discoveries of modern science coming upon us like bolts from the blue, opening our eyes to marvels we never dreamt of.

But many of these are only re-discoveries of what had been found ages ago. It was only the other day that modern science found that even in the midst of the variety of forces there is unity. It has just discovered that what it calls heat, magnetism, electricity, and so forth, are all convertible into one unit force, and as such, it expresses all these by one name, whatever you may choose to call it.

But this has been done even in the Samhita; old and ancient as it is, in it we meet with this very idea of force we were referring to. All the forces, whether you call them gravitation, or attraction, or repulsion, whether expressing themselves as heat, or electricity, or magnetism, are nothing but the variations of that unit energy. Whether they express themselves as thought, reflected from *Antahkarana*, the inner organs of man, or as action from an external organ, the unit from which they spring is what is called Prana. Again, what is Prana? Prana is Spandana or vibration.

When all this universe shall have resolved back into its primal state, what becomes of this infinite force? Do they think that it becomes extinct? Of course not. If it became extinct, what would be the cause of the next wave, because the motion is going in wave forms, rising, falling, rising again, falling again? Here is the word Srishti, which expresses the universe. Mark that the word does not mean creation. I am helpless in talking English; I have to translate the Sanskrit words as best as I can.

It is Srishti, projection. At the end of a cycle, everything becomes finer and finer and is resolved back into the primal state from which it sprang, and there it remains for a time quiescent, ready to spring forth again. That is Srishti, projection. And what becomes of all these forces, the Pranas? They are resolved back into the primal Prana, and this Prana becomes almost motionless—not entirely motionless; and that is what is described in the Vedic Sukta: "It vibrated without vibrations"—*Anidavaṭam*. There are many technical phrases in the Upanishads difficult to understand.

For instance, take this word Vata; many times it means air and many times motion, and often people confuse one with the other. We must guard against that. And what becomes of what you call matter? The forces permeate all matter; they all dissolve into *Akasha*, from which they again come out; this *Akasha* is the primal matter. Whether you translate it as ether or anything else, the idea is that this *Akasha* is the primal form of matter. This *Akasha* vibrates under the action of Prana, and when the next *Srishti* is coming up, as the vibration becomes quicker, the *Akasha* is lashed into all these wave forms which we call suns, moons, and systems.

We read again: "Everything in this universe has been projected, Prana vibrating." You must mark the word Ejati, because it comes from *Eja*—to vibrate. *Nihsritam*—projected. *Yadidam Kincha*—whatever in this universe.

This is a part of the cosmological side. There are many details working into it. For instance, how the process takes place, how there is first ether, and how from the ether come other things, how that ether begins to vibrate, and from that Vayu comes. But the one idea is here that it is from the finer that the grosser has come. Gross matter is the last to emerge and the most external, and this gross matter had the finer matter before it. Yet we see that the whole thing has been resolved into two, but there is not yet a final unity. There is the unity of force, Prana; there is the unity of matter, called *Akasha*. Is there any unity to be found among them again? Can they be melted into one? Our modern science is mute here, it has not yet found its way out; and if it is doing so, just as it has been slowly finding the same old Prana and the same ancient *Akasha*, it will have to move along the same lines.

The next unity is the omnipresent impersonal Being known by its old mythological name as Brahma, the four-headed Brahma, and psychologically called *Mahat*. This is where the two unite. What is called your mind is only a bit of this *Mahat* caught in the trap of the brain, and the sum total of all minds caught in the meshes of brains is what you call *Samashti*, the aggregate, the universal. Analysis had to go further; it was not yet complete. Here we were each one of us, as it were, a microcosm, and the world taken altogether is the macrocosm. But whatever is in the Vyashti, the particular, we may safely conjecture that a similar thing is happening also outside. If we had the power to analyse our own minds, we might safely conjecture that the same thing is happening in the cosmic mind. What is this mind is the question.

In modern times, in Western countries, as physical science is making rapid progress, as physiology is step by step conquering stronghold after stronghold of old religions, the Western people do not know where to stand, because to their great despair, modern physiology at every step has identified the mind with the brain. But we in India have known that always. That is the first proposition the Hindu boy learns that the mind is matter, only finer. The body is gross, and behind the body is what we call the Sukshma Sharira, the fine body, or mind. This is also material, only finer; and it is not the *Atman*.

I will not translate this word to you in English, because the idea does not exist in Europe; it is untranslatable. The modern attempt of German philosophers is to translate the word *Atman* by the word "Self", and until that word is universally accepted, it is impossible to use it. So, call it as Self or anything, it is our *Atman*. This *Atman* is the real man behind. It is the *Atman* that uses the material mind as its instrument, its *Antahkarana*, as is the psychological term for the mind. And the mind by means of a series of internal organs works the visible organs of the body. What is this mind?

It was only the other day that Western philosophers have come to know that the eyes are not the real organs of vision, but that behind these are other organs, the Indriyas, and if these are destroyed, a man may have a thousand eyes, like Indra, but there will be no sight for him. Ay, your philosophy starts with this assumption that by vision is not meant the external vision.

The real vision belongs to the internal organs, the brain-centres inside. You may call them what you like, but it is not that the Indriyas are the eyes, or the nose, or the ears. And the sum total of all these Indriyas plus the *Manas*, *Buddhi*, *Chitta*, *Ahamkara*, etc., is what is called the mind, and if the modern physiologist comes to tell you that the brain is what is called the mind, and that the brain is formed of so many organs, you need not be afraid at all; tell him that your philosophers knew it always; it is one of the very first principles of your religion.

Well then, we have to understand now what is meant by this *Manas*, *Buddhi*, *Chitta*, *Ahamkara*, etc. First of all, let us take *Chitta*. It is the mind-stuff—a part of the *Mahat*—it is the generic name for the mind itself, including all its various states. Suppose on a summer evening, there is a lake, smooth and calm, without a ripple on its surface. And suppose some one throws a stone into this lake.

What happens? First there is the action, the blow given to the water; next the water rises and sends a reaction towards the stone, and that reaction takes the form of a wave. First the water vibrates a little, and immediately sends back a reaction in the form of a wave. The *Chitta* let us compare to this lake, and the external objects are like the stones thrown into it. As soon as it comes in contact with any external object by means of these Indriyas—the Indriyas must be there to carry these external objects inside—there is a vibration, what is called *Manas*, indecisive. Next there is a reaction, the determinative faculty, *Buddhi*, and along with this *Buddhi* flashes the idea of Aham and the external object. Suppose there is a mosquito sitting upon my hand.

This sensation is carried to my *Chitta* and it vibrates a little; this is the psychological *Manas*. Then there is a reaction, and immediately comes the idea that I have a mosquito on my hand and that I shall have to drive it off. Thus these stones are thrown into the lake, but in the case of the lake every blow that comes to it is from the external world, while in the case of the lake of the mind, the blows may either come from the external world or the internal world. This whole series is what is called the *Antahkarana*.

Along with it, you ought to understand one thing more that will help us in understanding the Advaita system later on. It is this. All of you must have seen pearls and most of you know how pearls are formed. A grain of sand enters into the shell of a pearl oyster, and sets up an irritation there, and the oyster's body reacts towards the irritation and covers the little particle with its own juice.

That crystallises and forms the pearl. So the whole universe is like that, it is the pearl which is being formed by us. What we get from the external world is simply the blow. Even to be conscious of that blow we have to react, and as soon as we react, we really project a portion of our own mind towards the blow, and when we come to know of it, it is really our own mind as it has been shaped by the blow.

Therefore it is clear even to those who want to believe in a hard and fast realism of an external world, which they cannot but admit in these days of physiology—that supposing we represent the external world by "x", what we really know is "x" plus mind, and this mind-element is so great that it has covered the whole of that "x" which has remained unknown and unknowable throughout; and, therefore, if there is an external world, it is always unknown and unknowable. What we know of it is as it is moulded, formed, fashioned by our own mind.

So with the internal world. The same applies to our own soul, the *Atman*. In order to know the *Atman* we shall have to know It through the mind; and, therefore, what little we know of this *Atman* is simply the *Atman* plus the mind. That is to say, the *Atman* covered over, fashioned, and moulded by the mind, and nothing more. We shall return to this a little later, but we will remember what has been told here.

The next thing to understand is this. The question arose that this body is the name of one continuous stream of matter—every moment we are adding material to it, and every moment material is being thrown off by it—like a river continually flowing, vast masses of water always changing places; yet all the same, we take up the whole thing in imagination, and call it the same river. What do we call the river? Every moment the water is changing, the shore is changing, every moment the environment is changing, what is the river then?

It is the name of this series of changes. So with the mind. That is the great *Kshanika Vijnana Vada* doctrine, most difficult to understand, but most rigorously and logically worked out in the Buddhistic philosophy; and this arose in India in opposition to some part of the Vedanta. That had to be answered and we shall see later on how it could only be answered by Advaitism and by nothing else. We will see also how, in spite of people's curious notions about Advaitism, people's fright about Advaitism, it is the salvation of the world, because therein alone is to be found the reason of things. Dualism and other isms are very good as means of worship, very satisfying to the mind, and maybe, they have helped the mind onward; but if man wants to be rational and religious at the same time, Advaita is the one system in the world for him.

Well, now, we shall regard the mind as a similar river, continually filling itself at one end and emptying itself at the other end. Where is that unity which we call the *Atman*? The idea is this, that in spite of this continuous change in the body, and in spite of this continuous change in the mind, there is in us something that is unchangeable, which makes our ideas of things appear unchangeable. When rays of light coming from different quarters fall upon a screen, or a wall, or upon something that is not changeable, then and then alone it is possible for them to form a unity, then and then alone it is possible for them to form one complete whole.

Where is this unity in the human organs, falling upon which, as it were, the various ideas will come to unity and become one complete whole? This certainly cannot be the mind itself, seeing that it also changes. Therefore there must be something which is neither the body nor the mind, something which changes not, something permanent, upon which all our ideas, our sensations fall to form a unity and a complete whole; and this is the real soul, the *Atman* of man. And seeing that everything material, whether you call it fine matter, or mind, must be changeful, seeing that what you call gross matter, the external world, must also be changeful in comparison to that—this unchangeable something cannot be of material substance; therefore it is spiritual, that is to say, it is not matter—it is indestructible, unchangeable.

Next will come another question: Apart from those old arguments which only rise in the external world, the arguments in support of design—who created this external world, who created matter, etc.? The idea here is to know truth only from the inner nature of man, and the question arises just in the same way as it arose about the soul. Taking for granted that there is a soul, unchangeable, in each man, which is neither the mind nor the body, there is still a unity of idea among the souls, a unity of feeling, of sympathy. How is it possible that my soul can act upon your soul, where is the medium through which it can work, where is the medium through which it can act? How is it I can feel anything about your souls?

What is it that is in touch both with your soul and my soul? Therefore there is a metaphysical necessity of admitting another soul, for it must be a soul which acts in contact with all the different souls, and in and through matter—one Soul which covers and interpenetrates all the infinite number of souls in the world, in and through which they live, in and through which they sympathise, and love, and work for one another.

And this universal Soul is Paramatman, the Lord God of the universe. Again, it follows that because the soul is not made of matter, since it is spiritual, it cannot obey the laws of matter, it cannot be judged by the laws of matter. It is, therefore, unconquerable, birthless, deathless, and changeless.

"This Self, weapons cannot pierce, nor fire can burn, water cannot wet, nor air can dry up. Changeless, all-pervading, unmoving, immovable, eternal is this Self of man." We learn according to the Gita and the Vedanta that this individual Self is also Vibhu, and according to Kapila, is omnipresent. Of course there are sects in India which hold that the Self is Anu, infinitely small; but what they mean is Anu in manifestation; its real nature is Vibhu, all-pervading.

There comes another idea, startling perhaps, yet a characteristically Indian idea, and if there is any idea that is common to all our sects, it is this. Therefore I beg you to pay attention to this one idea and to remember it, for this is the very foundation of everything that we have in India. The idea is this. You have heard of the doctrine of physical evolution preached in the Western world by the German and the English savants.

It tells us that the bodies of the different animals are really one; the differences that we see are but different expressions of the same series; that from the lowest worm to the highest and the most saintly man it is but one—the one changing into the other, and so on, going up and up, higher and higher, until it attains perfection. We had that idea also. Declares our Yogi Patanjali—One species—the Jati is species—changes into another species—evolution; Pàrinama means one thing changing into another, just as one species changes into another. Where do we differ from the Europeans? Patanjali says, *"Prakrityapurat,"* by the infilling of nature.

The European says, it is competition, natural and sexual selection, etc. that forces one body to take the form of another. But here is another idea, a still better analysis, going deeper into the thing and saying, "By the infilling of nature". What is meant by this infilling of nature? We admit that the amoeba goes higher and higher until it becomes a Buddha; we admit that, but we are at the same time as much certain that you cannot get an amount of work out of a machine unless you have put it in some shape or other. The sum total of the energy remains the same, whatever the forms it may take.

If you want a mass of energy at one end, you have got to put it in at the other end; it may be in another form, but the amount of energy that should be produced out of it must be the same. Therefore, if a Buddha is the one end of the change, the very amoeba must have been the Buddha also. If the Buddha is the evolved amoeba, the amoeba was the involved Buddha also. If this universe is the manifestation of an almost infinite amount of energy, when this universe was in a state of Pralaya, it must have represented the same amount of involved energy.

It cannot have been otherwise. As such, it follows that every soul is infinite. From the lowest worm that crawls under our feet to the noblest and greatest saints, all have this infinite power, infinite purity, and infinite everything. Only the difference is in the degree of manifestation. The worm is only manifesting just a little bit of that energy, you have manifested more, another god-man has manifested still more: that is all the difference. But that infinite power is there all the same. Says Patanjali: "Like the peasant irrigating his field."

Through a little corner of his field he brings water from a reservoir somewhere, and perhaps he has got a little lock that prevents the water from rushing into his field. When he wants water, he has simply to open the lock, and in rushes the water of its own power. The power has not to be added, it is already there in the reservoir. So every one of us, every being, has as his own background such a reservoir of strength, infinite power, infinite purity, infinite bliss, and existence infinite—only these locks, these bodies, are hindering us from expressing what we really are to the fullest.

And as these bodies become more and more finely organised, as the Tamoguna becomes the Rajoguna, and as the Rajoguna becomes *Sattvaguna*, more and more of this power and purity becomes manifest, and therefore it is that our people have been so careful about eating and drinking, and the food question. It may be that the original ideas have been lost, just as with our marriage—which, though not belonging to the subject, I may take as an example. If we have another opportunity we will talk to you about these; but let me tell you now that the ideas behind our marriage system are the only ideas through which there can be a real civilisation. There cannot be anything else. If a man or a woman were allowed the freedom to take up any woman or man as wife or husband, if individual pleasure, satisfaction of animal instincts, were to be allowed to run loose in society, the result must be evil, evil children, wicked and demoniacal. The, man in every country is, on the one hand, producing these brutal children, and on the other hand multiplying the police force to keep these brutes down. The question is not how to destroy evil that way, but how to prevent the very birth of evil.

And so long as you live in society your marriage certainly affects every member of it; and therefore society has the right to dictate whom you shall marry, and whom you shall not. And great ideas of this kind have been behind the system of marriage here, what they call the astrological Jati of the bride and bridegroom. And in passing we may remark that according to Manu a child who is born of lust is not an Aryan. The child whose very conception and whose death is according to the rules of the Vedas, such is an Aryan. Yes, and less of these Aryan children are being produced in every country, and the result is the mass of evil which we call Kali Yuga.

But we have lost all these ideals—it is true we cannot carry all these ideas to the fullest length now—it is perfectly true we have made almost a caricature of some of these great ideas. It is lamentably true that the fathers and mothers are not what they were in old times, neither is society so educated as it used to be, neither has society that love for individuals that it used to have. But, however faulty the working out may be, the principle is sound; and if its application has become defective, if one method has failed, take up the principle and work it out better; why kill the principle? The same applies to the food question. The work and details are bad, very bad indeed, but that does not hurt the principle. The principle is eternal and must be there. Work it out afresh and make a re-formed application.

This is the one great idea of the *Atman* which every one of our sects in India has to believe. Only, as we shall find, the dualists preach that this *Atman* by evil works becomes *Sankuchita*, *i.e.* all its powers and its nature become contracted, and by good works again that nature expands. And the Advaitist says that the *Atman* never expands nor contracts, but seems to do so. It appears to have become contracted. That is all the difference, but all have the one idea that our *Atman* has all the powers already, not that anything will come to it from outside, not that anything will drop into it from the skies. Mark you, your Vedas are not inspired, but expired, not that they came from anywhere outside, but they are the eternal laws living in every soul.

The Vedas are in the soul of the ant, in the soul of the god. The ant has only to evolve and get the body of a sage or a Rishi, and the Vedas will come out, eternal laws expressing themselves. This is the one great idea to understand that our power is already ours, our salvation is already within us. Say either that it has become contracted, or say that it has been covered with the veil of Maya, it matters little; the idea is there already; you must have to believe in that, believe in the possibility of everybody—that even in the lowest man there is the same possibility as in the Buddha. This is the doctrine of the *Atman*.

But now comes a tremendous fight. Here are the Buddhists, who equally analyse the body into a material stream and as equally analyse the mind into another. And as for this *Atman*, they state that it is unnecessary; so we need not assume the *Atman* at all.

What use of a substance, and qualities adhering to the substance? We say, Gunas, qualities, and qualities alone. It is illogical to assume two causes where one will explain the whole thing. And the fight went on, and all the theories which held the doctrine of substance were thrown to the ground by the Buddhists. There was a break-up all along the line of those who held on to the doctrine of substance and qualities, that you have a soul, and I have a soul, and every one has a soul separate from the mind and body, and that each one is an individual.

So far we have seen that the idea of dualism is all right; for there is the body, there is then the fine body—the mind—there is this *Atman*, and in and through all the Atmans is that Paramatman, God. The difficulty is here that this *Atman* and Paramatman are both called substance, to which the mind and body and so-called substances adhere like so many qualities. Nobody has ever seen a substance, none can ever conceive; what is the use of thinking of this substance? Why not become a Kshanikavadin and say that whatever exists is this succession of mental currents and nothing more?

They do not adhere to each other, they do not form a unit, one is chasing the other, like waves in the ocean, never complete, never forming one unit-whole. Man is a succession of waves, and when one goes away it generates another, and the cessation of these wave-forms is what is called *Nirvana*. You see that dualism is mute before this; it is impossible that it can bring up any argument, and the dualistic God also cannot be retained here.

The idea of a God that is omnipresent, and yet is a person who creates without hands, and moves without feet, and so on, and who has created the universe as a *Kumbakara* (potter) creates a Ghata (pot), the Buddhist declares, is childish, and that if this is God, he is going to fight this God and not worship it. This universe is full of misery; if it is the work of a God, we are going to fight this God. And secondly, this God is illogical and impossible, as all of you are aware. We need not go into the defects of the "design theory", as all our *Kshanikas* have shown them full well; and so this Personal God fell to pieces.

Truth, and nothing but truth, is the watchword of the Advaitist. "Truth alone triumphs, and not untruth. Through truth alone the way to gods, Devayana, lies." Everybody marches forward under that banner; ay, but it is only to crush the weaker man's position by his own. You come with your dualistic idea of God to pick a quarrel with a poor man who is worshipping an image, and you think you are wonderfully rational, you can confound him; but if he turns round and shatters your own Personal God and calls that an imaginary ideal, where are you? You fall back on faith and so on, or raise the cry of atheism, the old cry of a weak man—whosoever defeats him is an atheist.

If you are to be rational, be rational all along the line, and if not, allow others the same privilege which you ask for yourselves. How can you prove the existence of this God? On the other hand, it can be almost disproved. There is not a shadow of a proof as to His existence, and there are very strong arguments to the contrary. How will you prove His existence, with your God, and His Gunas, and an infinite number of souls which are substance, and each soul an individual?

In what are you an individual? You are not as a body, for you know today better than even the Buddhists of old knew that what may have been matter in the sun has just now become matter in you, and will go out and become matter in the plants; then where is your individuality, Mr. So-and-so? The same applies to the mind. Where is your individuality? You have one thought tonight and another tomorrow. You do not think the same way as you thought when you were a child; and old men do not think the same way as they did when they were young.

Where is your individuality then? Do not say it is in consciousness, this *Ahamkara*, because this only covers a small part of your existence. While I am talking to you, all my organs are working and I am not conscious of it. If consciousness is the proof of existence they do not exist then, because I am not conscious of them. Where are you then with your Personal God theories? How can you prove such a God?

Again, the Buddhists will stand up and declare—not only is it illogical, but immoral, for it teaches man to be a coward and to seek assistance outside, and nobody can give him such help. Here is the universe, man made it; why then depend on an imaginary being outside whom nobody ever saw, or felt, or got help from? Why then do you make cowards of yourselves and teach your children that the highest state of man is to be like a dog, and go crawling before this imaginary being, saying that you are weak and impure, and that you are everything vile in this universe?

On the other hand, the Buddhists may urge not only that you tell a lie, but that you bring a tremendous amount of evil upon your children; for, mark you, this world is one of hypnotisation. Whatever you tell yourself, that you become. Almost the first words the great Buddha uttered were: "What you think, that you are; what you will think, that you will be." If this is true, do not teach yourself that you are nothing, ay, that you cannot do anything unless you are helped by somebody who does not live here, but sits above the clouds. The result will be that you will be more and more weakened every day. By constantly repeating, "we are very impure, Lord, make us pure", the result will be that you will hypnotise yourselves into all sorts of vices.

The Buddhists say that ninety per cent of these vices that you see in every society are on account of this idea of a Personal God; this is an awful idea of the human being that the end and aim of this expression of life, this wonderful expression of life, is to become like a dog. Says the Buddhist to the Vaishnava, if your ideal, your aim and goal is to go to the place called Vaikuntha where God lives, and there stand before Him with folded hands all through eternity, it is better to commit suicide than do that. The Buddhists may even urge that, that is why he is going to create annihilation, *Nirvana*, to escape this. I am putting these ideas before you as a Buddhist just for the time being, because nowadays all these Advaitic ideas are said to make you immoral, and I am trying to tell you how the other side looks. Let us face both sides boldly and bravely.

We have seen first of all that this cannot be proved, this idea of a Personal God creating the world; is there any child that can believe this today? Because a Kumbhara creates a Ghata, therefore a God created the world! If this is so, then your Kumbhara is God also; and if any one tells you that He acts without head and hands, you may take him to a lunatic asylum. Has ever your Personal God, the Creator of the world to whom you cry all your life, helped you—is the next challenge from modern science.

They will prove that any help you have had could have been got by your own exertions, and better still, you need not have spent your energy in that crying, you could have done it better without that weeping and crying. And we have seen that along with this idea of a Personal God comes tyranny and priest-craft. Tyranny and priest-craft have prevailed wherever this idea existed, and until the lie is knocked on the head, say the Buddhists, tyranny will not cease. So long as man thinks he has to cower before a supernatural being, so long there will be priests to claim rights and privileges and to make men cower before them, while these poor men will continue to ask some priest to act as interceder for them. You may do away with the Brahmin, but mark me, those who do so will put themselves in his place and will be worse, because the Brahmin has a certain amount of generosity in him, but these upstarts are always the worst of tyrannisers.

If a beggar gets wealth, he thinks the whole world is a bit of straw. So these priests there must be, so long as this Personal God idea persists, and it will be impossible to think of any great morality in society. Priestcraft and tyranny go hand in hand. Why was it invented? Because some strong men in old times got people into their hands and said, you must obey us or we will destroy you. That was the long and short of it. It is the idea of the thunderer who kills every one who does not obey him. Next the Buddhist says, you have been perfectly rational up to this point, that everything is the result of the law of Karma. You believe in an infinity of souls, and that souls are without birth or death, and this infinity of souls and the belief in the law of Karma are perfectly logical no doubt. There cannot be a cause without an effect, the present must have had its cause in the past and will have its effect in the future.

The Hindu says the Karma is Jada (inert) and not Chaitanya (Spirit), therefore some Chaitanya is necessary to bring this cause to fruition. Is it so, that Chaitanya is necessary to bring the plant to fruition? If I plant the seed and add water, no Chaitanya is necessary. You may say there was some original Chaitanya there, but the souls themselves were the Chaitanya, nothing else is necessary. If human souls have it too, what necessity is there for a God, as say the Jains, who, unlike the Buddhists, believe in souls and do not believe in God. Where are you logical, where are you moral? And when you criticise Advaitism and fear that it will make for immorality, just read a little of what has been done in India by dualistic sects.

If there have been twenty thousand Advaitist blackguards, there have also been twenty thousand Dvaitist blackguards. Generally speaking, there will be more Dvaitist blackguards, because it takes a better type of mind to understand Advaitism, and Advaitists can scarcely be frightened into anything. What remains for you Hindus, then? There is no help for you out of the clutches of the Buddhists. You may quote the Vedas, but he does not believe in them. He will say, "My Tripitakas say otherwise, and they are without beginning or end, not even written by Buddha, for Buddha says he is only reciting them; they are eternal." And he adds, "Yours are wrong, ours are the true Vedas, yours are manufactured by the Brahmin priests, therefore out with them." How do you escape?

Here is the way to get out. Take up the first objection, the metaphysical one, that substance and qualities are different. Says the Advaitist, they are not. There is no difference between substance and qualities. You know the old illustration, how the rope is taken for the snake, and when you see the snake you do not see the rope at all, the rope has vanished. Dividing the thing into substance and quality is a metaphysical something in the brains of philosophers, for never can they be in effect outside. You see qualities if you are an ordinary man, and substance if you are a great Yogi, but you never see both at the same time.

So, Buddhists, your quarrel about substance and qualities has been but a miscalculation which does not stand on fact. But if substance is unqualified, there can only be one. If you take qualities off from the soul, and show that these qualities are in the mind really, superimposed on the soul, then there can never be two souls for it is qualification that makes the difference between one soul and another. How do you know that one soul is different from the other? Owing to certain differentiating marks, certain qualities. And where qualities do not exist, how can there be differentiation? Therefore there are not two souls, there is but One, and your Paramatman is unnecessary, it is this very soul. That One is called Paramatman, that very One is called Jivatman, and so on; and you dualists, such as the Sankhyas and others, who say that the soul is Vibhu, omnipresent, how can you make two infinites? There can be only one. What else? This One is the one Infinite *Atman*, everything else is its manifestation. There the Buddhist stops, but there it does not end.

The Advaitist position is not merely a weak one of criticism. The Advaitist criticises others when they come too near him, and just throws them away, that is all; but he propounds his own position. He is the only one that criticises, and does not stop with criticism and showing books. Here you are. You say the universe is a thing of continuous motion. In Vyashti (the finite) everything is moving; you are moving, the table is moving, motion everywhere; it is Samsara, continuous motion; it is Jagat. Therefore there cannot be an individuality in this Jagat, because individuality means that which does not change; there cannot be any changeful individuality, it is a contradiction in terms. There is no such thing as individuality in this little world of ours, the Jagat. Thought and feeling, mind and body, men and animals and plants are in a continuous state of flux. But suppose you take the universe as a unit whole; can it change or move? Certainly not.

Motion is possible in comparison with something which is a little less in motion or entirely motionless. The universe as a whole, therefore, is motionless, unchangeable. You are therefore, an individual then and then alone when you are the whole of it, when the realisation of "I am the universe" comes. That is why the Vedantist says that so long as there are two, fear does not cease. It is only when one does not see another, does not feel another, when it is all one—then alone fear ceases, then alone death vanishes, then alone Samsara vanishes. Advaita teaches us, therefore, that man is individual in being universal, and not in being particular.

You are immortal only when you are the whole. You are fearless and deathless only when you are the universe; and then that which you call the universe is the same as that you call God, the same that you call existence, the same that you call the whole. It is the one undivided Existence which is taken to be the manifold world which we see, as also others who are in the same state of mind as we. People who have done a little better Karma and get a better state of mind, when they die, look upon it as *Svarga* and see Indras and so forth. People still higher will see it, the very same thing, as Brahma-Loka, and the perfect ones will neither see the earth nor the heavens, nor any Loka at all. The universe will have vanished, and Brahman will be in its stead.

Can we know this Brahman? I have told you of the painting of the Infinite in the Samhita. Here we shall find another side shown, the infinite internal. That was the infinite of the muscles. Here we shall have the Infinite of thought. There the Infinite was attempted to be painted in language positive; here that language failed and the attempt has been to paint it in language negative. Here is this universe, and even admitting that it is Brahman, can we know it? No! No! You must understand this one thing again very clearly. Again and again this doubt will come to you: If this is Brahman, how can we know it? "By what can the knower be known?" How can the knower be known? The eyes see everything; can they see themselves? They cannot. The very fact of knowledge is a degradation. Children of the Aryans, you must remember this, for herein lies a big story. All the Western temptations that come to you, have their metaphysical basis on that one thing—there is nothing higher than sense-knowledge.

In the East, we say in our Vedas that this knowledge is lower than the thing itself, because it is always a limitation. When you want to know a thing, it immediately becomes limited by your mind. They say, refer back to that instance of the oyster making a pearl and see how knowledge is limitation, gathering a thing, bringing it into consciousness, and not knowing it as a whole. This is true about all knowledge, and can it be less so about the Infinite? Can you thus limit Him who is the substance of all knowledge, Him who is the Sakshi, the witness, without whom you cannot have any knowledge, Him who has no qualities, who is the Witness of the whole universe, the Witness in our own souls? How can you know Him? By what means can you bind Him up? Everything, the whole universe, is such a false attempt. This infinite *Atman* is, as it were, trying to see His own face, and all, from the lowest animals to the highest of gods, are like so many mirrors to reflect Himself in, and He is taking up still others, finding them insufficient, until in the human body He comes to know that it is the finite of the finite, all is finite, there cannot be any expression of the Infinite in the finite.

Then comes the retrograde march, and this is what is called renunciation, Vairagya. Back from the senses, back! Do not go to the senses is the watchword of Vairagya. This is the watchword of all morality, this is the watchword of all well-being; for you must remember that with us the universe begins in Tapasya, in renunciation, and as you go back and back, all the forms are being manifested before you, and they are left aside one after the other until you remain what you really are. This is Moksha or liberation.

This idea we have to understand: "How to know the knower?" The knower cannot be known, because if it were known, it will not be the knower. If you look at your eyes in a mirror, the reflection is no more your eyes, but something else, only a reflection. Then if this Soul, this Universal, Infinite Being which you are, is only a witness, what good is it? It cannot live, and move about, and enjoy the world, as we do. People cannot understand how the witness can enjoy. "Oh," they say, "you Hindus have become quiescent, and good for nothing, through this doctrine that you are witnesses!" First of all, it is only the witness that can enjoy.

If there is a wrestling match, who enjoys it, those who take part in it, or those who are looking on—the outsiders? The more and more you are the witness of anything in life, the more you enjoy it. And this is Ananda; and, therefore, infinite bliss can only be yours when you have become the witness of this universe; then alone you are a Mukta Purusha. It is the witness alone that can work without any desire, without any idea of going to heaven, without any idea of blame, without any idea of praise. The witness alone enjoys, and none else.

Coming to the moral aspect, there is one thing between the metaphysical and the moral aspect of Advaitism; it is the theory of Maya. Everyone of these points in the Advaita system requires years to understand and months to explain. Therefore you will excuse me if I only just touch them en passant. This theory of Maya has been the most difficult thing to understand in all ages. Let me tell you in a few words that it is surely no theory, it is the combination of the three ideas *Desha-Kala-Nimitta*—space, time, and causation—and this time and space and cause have been further reduced into Nama-Rupa. Suppose there is a wave in the ocean.

The wave is distinct from the ocean only in its form and name, and this form and this name cannot have any separate existence from the wave; they exist only with the wave. The wave may subside, but the same amount of water remains, even if the name and form that were on the wave vanish for ever. So this Maya is what makes the difference between me and you, between all animals and man, between gods and men. In fact, it is this Maya that causes the *Atman* to be caught, as it were, in so many millions of beings, and these are distinguishable only through name and form. If you leave it alone, let name and form go, all this variety vanishes for ever, and you are what you really are. This is Maya.

It is again no theory, but a statement of facts. When the realist states that this table exists, what he means is, that this table has an independent existence of its own, that it does not depend on the existence of anything else in the universe, and if this whole universe be destroyed and annihilated, this table will remain just as it is now. A little thought will show you that it cannot be so.

Everything here in the sense-world is dependent and interdependent, relative and correlative, the existence of one depending on the other. There are three steps, therefore, in our knowledge of things; the first is that each thing is individual and separate from every other; and the next step is to find that there is a relation and correlation between all things; and the third is that there is only one thing which we see as many.

The first idea of God with the ignorant is that this God is somewhere outside the universe, that is to say, the conception of God is extremely human; He does just what a man does, only on a bigger and higher scale. And we have seen how that idea of God is proved in a few words to be unreasonable and insufficient. And the next idea is the idea of a power we see manifested everywhere.

This is the real Personal God we get in the Chandi, but, mark me, not a God that you make the reservoir of all good qualities only. You cannot have two Gods, God and Satan; you must have only one and dare to call Him good and bad. Have only one and take the logical consequences. We read in the Chandi: "We salute Thee, O Divine Mother, who lives in every being as peace.

We salute Thee, O Divine Mother, who lives in all beings as purity." At the same time we must take the whole consequence of calling Him the All-formed. "All this is bliss, O Gargi; wherever there is bliss there is a portion of the Divine." You may use it how you like. In this light before me, you may give a poor man a hundred rupees, and another man may forge your name, but the light will be the same for both. This is the second stage. And the third is that God is neither outside nature nor inside nature, but God and nature and soul and universe are all convertible terms.

You never see two things; it is your metaphysical words that have deluded you. You assume that you are a body and have a soul, and that you are both together. How can that be? Try in your own mind. If there is a Yogi among you, he knows himself as Chaitanya, for him the body has vanished. An ordinary man thinks of himself as a body; the idea of spirit has vanished from him; but because the metaphysical ideas exist that man has a body and a soul and all these things, you think they are all simultaneously there. One thing at a time. Do not talk of God when you see matter; you see the effect and the effect alone, and the cause you cannot see, and the moment you can see the cause, the effect will have vanished. Where is the world then, and who has taken it off?

"One that is present always as consciousness, the bliss absolute, beyond all bounds, beyond all compare, beyond all qualities, ever-free, limitless as the sky, without parts, the absolute, the perfect—such a Brahman, O sage, O learned one, shines in the heart of the Jnani in Samadhi. (Vivekachudamani, 408)

"Where all the changes of nature cease for ever, who is thought beyond all thoughts, who is equal to all yet having no equal, immeasurable, whom the Vedas declare, who is the essence in what we call our existence, the perfect—such a Brahman, O Sage, O learned one, shines in the heart of the Jnani in Samadhi. (Ibid.,409)

"Beyond all birth and death, the Infinite One, incomparable, like the whole universe deluged in water in Mahapralaya—water above, water beneath, water on all sides, and on the face of that water not a wave, not a ripple—silent and calm, all visions have died out, all fights and quarrels and the war of fools and saints have ceased for ever—such a Brahman, O sage, O learned one, shines in the heart of the Jnani in Samadhi." (Ibid., 410)

That also comes, and when that comes the world has vanished. We have seen then that this Brahman, this Reality is unknown and unknowable, not in the sense of the agnostic, but because to know Him would be a blasphemy, because you are He already. We have also seen that this Brahman is not this table and yet is this table. Take off the name and form, and whatever is reality is He. He is the reality in everything.

"Thou art the woman, thou the man, thou art the boy, and the girl as well, thou the old man supporting thyself on a stick, thou art all in all in the universe." That is the theme of Advaitism. A few words more. Herein lies, we find, the explanation of the essence of things. We have seen how here alone we can take a firm stand against all the onrush of logic and scientific knowledge. Here at least reason has a firm foundation, and, at the same time, the Indian Vedantist does not curse the preceding steps; he looks back and he blesses them, and he knows that they were true, only wrongly perceived, and wrongly stated.

They were the same truth, only seen through the glass of Maya, distorted it may be—yet truth, and nothing but truth. The same God whom the ignorant man saw outside nature, the same whom the little-knowing man saw as interpenetrating the universe, and the same whom the sage realises as his own Self, as the whole universe itself—all are One and the same Being, the same entity seen from different standpoints, seen through different glasses of Maya, perceived by different minds, and all the difference was caused by that. Not only so, but one view must lead to the other.

What is the difference between science and common knowledge? Go out into the streets in the dark, and if something unusual is happening there, ask one of the passers-by what is the cause of it. It is ten to one that he will tell you it is a ghost causing the phenomenon. He is always going after ghosts and spirits outside, because it is the nature of ignorance to seek for causes outside of effects. If a stone falls, it has been thrown by a devil or a ghost, says the ignorant man, but the scientific man says it is the law of nature, the law of gravitation.

What is the fight between science and religion everywhere? Religions are encumbered with such a mass of explanations which come from outside—one angel is in charge of the sun, another of the moon, and so on ad infinitum. Every change is caused by a spirit, the one common point of agreement being that they are all outside the thing. Science means that the cause of a thing is sought out by the nature of the thing itself. As step by step science is progressing, it has taken the explanation of natural phenomena out of the hands of spirits and angels. Because Advaitism has done likewise in spiritual matters, it is the most scientific religion. This universe has not been created by any extra-cosmic God, nor is it the work of any outside genius. It is self-creating, self-dissolving, self-manifesting, One Infinite Existence, the Brahman. *Tattvamasi Shvetaketo*—"That thou art, O Shvetaketu!"

Thus you see that this, and this alone, and none else, can be the only scientific religion. And with all the prattle about science that is going on daily at the present time in modern half-educated India, with all the talk about rationalism and reason that I hear every day, I expect that whole sects of you will come over and dare to be Advaitists, and dare to preach it to the world in the words of Buddha, "For the good of many, for the happiness of many." If you do not, I take you for cowards. If you cannot get over your cowardice, if your fear is your excuse, allow the same liberty to others, do not try to break up the poor idol-worshipper, do not call him a devil, do not go about preaching to every man that does not agree entirely with you. Know first, that you are cowards yourselves, and if society frightens you, if your own superstitions of the past frighten you so much, how much more will these superstitions frighten and bind down those who are ignorant? That is the

Advaita position. Have mercy on others. Would to God that the whole world were Advaitists tomorrow, not only in theory, but in realisation. But if that cannot be, let us do the next best thing; let us take the ignorant by the hand, lead them always step by step just as they can go, and know that every step in all religious growth in India has been progressive. It is not from bad to good, but from good to better.

Something more has to be told about the moral relation. Our boys blithely talk nowadays; they learn from somebody—the Lord knows from whom—that Advaita makes people immoral, because if we are all one and all God, what need of morality will there be at all! In the first place, that is the argument of the brute, who can only be kept down by the whip. If you are such brutes, commit suicide rather than pass for human beings who have to kept down by the whip. If the whip is taken away, you will all be demons! You ought all to be killed if such is the case. There is no help for you; you must always be living under this whip and rod, and there is no salvation, no escape for you.

In the second place, Advaita and Advaita alone explains morality. Every religion preaches that the essence of all morality is to do good to others. And why? Be unselfish. And why should I? Some God has said it? He is not for me. Some texts have declared it? Let them; that is nothing to me; let them all tell it. And if they do, what is it to me? Each one for himself, and somebody take the hindermost—that is all the morality in the world, at least with many. What is the reason that I should be moral? You cannot explain it except when you come to know the truth as given in the Gita: "He who sees everyone in himself, and himself in everyone, thus seeing the same God living in all, he, the sage, no more kills the Self by the self." Know through Advaita that whomsoever you hurt, you hurt yourself; they are all you.

Whether you know it or not, through all hands you work, through all feet you move, you are the king enjoying in the palace, you are the beggar leading that miserable existence in the street; you are in the ignorant as well as in the learned, you are in the man who is weak, and you are in the strong;

know this and be sympathetic. And that is why we must not hurt others. That is why I do not even care whether I have to starve, because there will be millions of mouths eating at the same time, and they are all mine. Therefore I should not care what becomes of me and mine, for the whole universe is mine, I am enjoying all the bliss at the same time; and who can kill me or the universe? Herein is morality. Here, in Advaita alone, is morality explained. The others teach it, but cannot give you its reason. Then, so far about explanation.

What is the gain? It is strength. Take off that veil of hypnotism which you have cast upon the world, send not out thoughts and words of weakness unto humanity. Know that all sins and all evils can be summed up in that one word, weakness. It is weakness that is the motive power in all evil doing; it is weakness that is the source of all selfishness; it is weakness that makes men injure others; it is weakness that makes them manifest what they are not in reality. Let them all know what they are; let them repeat day and night what they are. Soham. Let them suck it in with their mothers' milk, this idea of strength—I am He, I am He. This is to be heard first, etc. And then let them think of it, and out of that thought, out of that heart will proceed works such as the world has never seen. What has to be done? And, this Advaita is said by some to be impracticable; that is to say, it is not yet manifesting itself on the material plane. To a certain extent that is true, for remember the saying of the Vedas:

"Om, this is the Brahman; Om, this is the greatest reality; he who knows the secret of this Om, whatever he desires that he gets." Ay, therefore first know the secret of this Om, that you are the Om; know the secret of this *Tattvamasi*, and then and then alone whatever you want shall come to you. If you want to be great materially, believe that you are so. I may be a little bubble, and you may be a wave mountain-high, but know that for both of us the infinite ocean is the background, the infinite Brahman is our magazine of power and strength, and we can draw as much as we like, both of us, I the bubble and you the mountain-high wave. Believe, therefore, in yourselves. The secret of Advaita is: Believe in yourselves first, and then believe in anything else. In the history of the world, you will find that only those nations that have believed in themselves have become great and strong.

In the history of each nation, you will always find that only those individuals who have believed in themselves have become great and strong. Here, to India, came an Englishman who was only a clerk, and for want of funds and other reasons he twice tried to blow his brains out; and when he failed, he believed in himself, he believed that he was born to do great things; and that man became Lord Clive, the founder of the Empire. If he had believed the *Padres* and gone crawling all his life—"O Lord, I am weak, and I am low"—where would he have been? In a lunatic asylum. You also are made lunatics by these evil teachings. I have seen, all the world over, the bad effects of these weak teachings of humility destroying the human race. Our children are brought up in this way, and is it a wonder that they become semi-lunatics?

This is teaching on the practical side. Believe, therefore, in yourselves, and if you want material wealth, work it out; it will come to you. If you want to be intellectual, work it out on the intellectual plane, and intellectual giants you shall be. And if you want to attain to freedom, work it out on the spiritual plane, and free you shall be and shall enter into *Nirvana*, the Eternal Bliss. But one defect which lay in the Advaita was its being worked out so long on the spiritual plane only, and nowhere else; now the time has come when you have to make it practical.

It shall no more be a Rahasya, a secret, it shall no more live with monks in cave and forests, and in the Himalayas; it must come down to the daily, everyday life of the people; it shall be worked out in the palace of the king, in the cave of the recluse; it shall be worked out in the cottage of the poor, by the beggar in the street, everywhere; anywhere it can be worked out. Therefore do not fear whether you are a woman or a Shudra, for this religion is so great, says Lord Krishna, that even a little of it brings a great amount of good. Therefore, children of the Aryans, do not sit idle; awake, arise, and stop not till the goal is reached. The time has come when this

Advaita is to be worked out practically. Let us bring it down from heaven unto the earth; this is the present dispensation. And, the voices of our forefathers of old are telling us to bring it down from heaven to the earth. Let your teachings permeate the world, till they have entered into every pore of society, till they have become the common property of everybody, till they have become part and parcel of our lives, till they have entered into our veins and tingle with every drop of blood there.

You may be astonished to hear that as practical Vedantists the Americans are better than we are. I used to stand on the seashore at New York and look at the immigrants coming from different countries—crushed, downtrodden, hopeless, unable to look a man in the face, with a little bundle of clothes as all their possession, and these all in rags; if they saw a policeman they were afraid and tried to get to the other side of the foot-path. And, mark you, in six months those very men were walking erect, well clothed, looking everybody in the face; and what made this wonderful difference? Say, this man comes from Armenia or somewhere else where he was crushed down beyond all recognition, where everybody told him he was a born slave and born to remain in a low state all his life, and where at the least move on his part he was trodden upon.

There everything told him, as it were, "Slave! you are a slave, remain so. Hopeless you were born, hopeless you must remain." Even the very air murmured round him, as it were, "There is no hope for you; hopeless and a slave you must remain", while the strong man crushed the life out of him. And when he landed in the streets of New York, he found a gentleman, well-dressed, shaking him by the hand; it made no difference that the one was in rags and the other well-clad. He went a step further and saw a restaurant, that there were gentlemen dining at a table, and he was asked to take a seat at the corner of the same table.

He went about and found a new life, that there was a place where he was a man among men. Perhaps he went to Washington, shook hands with the President of the United States, and perhaps there he saw men coming from distant villages, peasants, and ill-clad, all shaking hands with the

President. Then the veil of Maya slipped away from him. He is Brahman, he who has been hypnotised into slavery and weakness is once more awake, and he rises up and finds himself a man in a world of men.

In this country of ours, the very birth-place of the Vedanta, our masses have been hypnotised for ages into that state. To touch them is pollution, to sit with them is pollution! Hopeless they were born, hopeless they must remain! And the result is that they have been sinking, sinking, sinking, and have come to the last stage to which a human being can come. For what country is there in the world where man has to sleep with the cattle? And for this, blame nobody else, do not commit the mistake of the ignorant. The effect is here and the cause is here too. We are to blame. Stand up, be bold, and take the blame on your own shoulders. Do not go about throwing mud at others; for all the faults you suffer from, you are the sole and only cause.

Young men of Lahore, understand this, therefore, this great sin, hereditary and national, is on our shoulders. There is no hope for us. You may make thousands of societies, twenty thousand political assemblages, fifty thousand institutions. These will be of no use until there is that sympathy, that love, that heart that thinks for all; until Buddha's heart comes once more into India, until the words of the Lord Krishna are brought to their practical use, there is no hope for us. You go on imitating the Europeans and their societies and their assemblages, but let me tell you a story, a fact that I saw with my own eyes.

A company of Burmans was taken over to London by some persons here, who turned out to be Eurasians. They exhibited these people in London, took all the money, and then took these Burmans over to the Continent, and left them there for good or evil. These poor people did not know a word of any European language, but the English Consul in Austria sent them over to London. They were helpless in London, without knowing anyone. But an English lady got to know of them, took these foreigners from Burma into her own house, gave them her own clothes, her bed, and everything, and then sent the news to the papers. And, mark you, the next day the whole nation was, as it were, roused. Money poured in, and these people were helped out and sent back to Burma.

On this sort of sympathy are based all their political and other institutions; it is the rock-foundation of love, for themselves at least. They may not love the world; and the Burmans may be their enemies, but in England, it goes without saying, there is this great love for their own people, for truth and justice and charity to the stranger at the door. Where is the heart here to build upon? No sooner do we start a little joint-stock company than we try to cheat each other, and the whole thing comes down with a crash. You talk of imitating the English and building up as big a nation as they are. But where are the foundations? Ours are only sand, and, therefore, the building comes down with a crash in no time.

Therefore, young men of Lahore, raise once more that mighty banner of Advaita, for on no other ground can you have that wonderful love until you see that the same Lord is present everywhere. Unfurl that banner of love! "Arise, awake, and stop not till the goal is reached." Arise, arise once more, for nothing can be done without renunciation. If you want to help others, your little self must go. In the words of the Christians—you cannot serve God and Mammon at the same time. Have Vairagya. Your ancestors gave up the world for doing great things.

At the present time there are men who give up the world to help their own salvation. Throw away everything, even your own salvation, and go and help others. Ay, you are always talking bold words, but here is practical Vedanta before you. Give up this little life of yours. What matters it if you die of starvation—you and I and thousands like us—so long as this nation lives? The nation is sinking, the curse of unnumbered millions is on our heads—those to whom we have been giving ditch-water to drink when they have been dying of thirst and while the perennial river of water was flowing past, the unnumbered millions whom we have allowed to starve in sight of plenty, the unnumbered millions to whom we have talked of Advaita and whom we have hated with all our strength, the unnumbered millions for whom we have invented the doctrine of Lokachara (usage), to whom we have talked theoretically that we are all the same and all are one with the same Lord, without even an ounce of practice.

"Yet, my friends, it must be only in the mind and never in practice!" Wipe off this blot. "Arise and awake." What matters it if this little life goes? Everyone has to die, the saint or the sinner, the rich or the poor. The body never remains for anyone. Arise and awake and be perfectly sincere. Our insincerity in India is awful; what we want is character, that steadiness and character that make a man cling on to a thing like grim death. "Let the sages blame or let them praise, let Lakshmi come today or let her go away, let death come just now or in a hundred years; he indeed is the sage who does not make one false step from the right path." Arise and awake, for the time is passing and all our energies will be frittered away in vain talking. Arise and awake, let minor things, and quarrels over little details and fights over little doctrines be thrown aside, for here is the greatest of all works, here are the sinking millions. When the Mohammedans first came into India, what a great number of Hindus were here; but mark, how today they have dwindled down! Every day they will become less and less till they wholly disappear.

Let them disappear, but with them will disappear the marvellous ideas, of which, with all their defects and all their misrepresentations, they still stand as representatives. And with them will disappear this marvellous Advaita, the crest-jewel of all spiritual thought. Therefore, arise, awake, with your hands stretched out to protect the spirituality of the world. And first of all, work it out for your own country. What we want is not so much spirituality as a little of the bringing down of the Advaita into the material world.

First bread and then religion. We stuff them too much with religion, when the poor fellows have been starving. No dogmas will satisfy the cravings of hunger. There are two curses here: first our weakness, secondly, our hatred, our dried-up hearts. You may talk doctrines by the millions, you may have sects by the hundreds of millions; ay, but it is nothing until you have the heart to feel. Feel for them as your Veda teaches you, till you find they are parts of your own bodies, till you realise that you and they, the poor and the rich, the saint and the sinner, are all parts of One Infinite Whole, which you call Brahman.

Modern science and its sledge-hammer blows are pulverising the porcelain foundations of all dualistic religions everywhere. Not only here are the dualists torturing texts till they will extend no longer—for texts are not India-rubber—it is not only here that they are trying to get into the nooks and corners to protect themselves; it is still more so in Europe and America.

And even there something of this idea will have to go from India. It has already got there. It will have to grow and increase and save their civilisations too. For in the West the old order of things is vanishing, giving way to a new order of things, which is the worship of gold, the worship of Mammon.

Thus this old crude system of religion was better than the modern system, namely—competition and gold. No nation, however strong, can stand on such foundations, and the history of the world tells us that all that had such foundations are dead and gone. In the first place we have to stop the incoming of such a wave in India.

Therefore preach the Advaita to every one, so that religion may withstand the shock of modern science. Not only so, you will have to help others; your thought will help out Europe and America. But above all, let me once more remind you that here is need of practical work, and the first part of that is that you should go to the sinking millions of India, and take them by the hand, remembering the words of the Lord Krishna:

"Even in this life they have conquered relative existence whose minds are firm-fixed on the sameness of everything, for God is pure and the same to all; therefore, such are said to be living in God."

Practical Vedanta 1

A very renoun scholar and historian has presented his views and exkperience about the Vedanta in the following pages:

If we have to say something about the practical position of the Vedanta philosophy. As we have told you, theory is very good indeed, but how are we to carry it into practice? If it be absolutely impracticable, no theory is of any value whatever, except as intellectual gymnastics. The Vedanta, therefore, as a religion must be intensely practical. We must be able to carry it out in every part of our lives. And not only this, the fictitious differentiation between religion and the life of the world must vanish, for the Vedanta teaches oneness—one life throughout.

The ideals of religion must cover the whole field of life, they must enter into all our thoughts, and more and more into practice. We will enter gradually on the practical side as we proceed. But this series of lectures is intended to be a basis, and so we must first apply ourselves to theories and understand how they are worked out, proceeding from forest caves to busy streets and cities; and one peculiar feature we find is that many of these thoughts have been the outcome, not of retirement into forests, but have emanated from persons whom we expect to lead the busiest lives—from ruling monarchs.

Shvetaketu was the son of Aruni, a sage, most probably a recluse. He was brought up in the forest, but he went to the city of the Panchalas and appeared at the court of the king, Pravahana Jaivali. The king asked him, "Do you know how beings depart hence at death?" "No, sir." "Do you know how they return hither?" "No, sir." "Do you know the way of the fathers and the way of the gods?" "No, sir." Then the king asked other questions. *Shvetaketu* could not answer them. So the king told him that he knew nothing. The boy went back to his father, and the father admitted that he himself could not answer these questions. It was not that he was unwilling to answer these questions.

It was not that he was unwilling to teach the boy, but he did not know these things. So he went to the king and asked to be taught these secrets. The king said that these things had been hitherto known only among kings; the priests never knew them. He, however, proceeded to teach him what he desired to know. In various Upanishads we find that this Vedanta philosophy is not the outcome of meditation in the forests only, but that the very best parts of it were thought out and expressed by brains which were busiest in the everyday affairs of life. We cannot conceive any man busier than an absolute monarch, a man who is ruling over millions of people, and yet, some of these rulers were deep thinkers.

Everything goes to show that this philosophy must be very practical; and later on, when we come to the Bhagavad-Gita—most of you, perhaps, have read it, it is the best commentary we have on the Vedanta philosophy—curiously enough the scene is laid on the battlefield, where Krishna teaches this philosophy to Arjuna; and the doctrine which stands out luminously in every page of the Gita is intense activity, but in the midst of it, eternal calmness. This is the secret of work, to attain which is the goal of the Vedanta. Inactivity, as we understand it in the sense of passivity, certainly cannot be the goal. Were it so, then the walls around us would be the most intelligent; they are inactive. Clods of earth, stumps of trees, would be the greatest sages in the world; they are inactive. Nor does inactivity become activity when it is combined with passion. Real activity, which is the goal of Vedanta, is combined with eternal calmness, the calmness which cannot be ruffled, the balance of mind which is never disturbed, whatever happens. And we all know from our experience in life that that is the best attitude for work.

I have been asked many times how we can work if we do not have the passion which we generally feel for work. I also thought in that way years ago, but as I am growing older, getting more experience, I find it is not true. The less passion there is, the better we work. The calmer we are, the better for us, and the more the amount of work we can do. When we let loose our feelings, we waste so much energy, shatter our nerves, disturb our minds, and accomplish very little work.

The energy which ought to have gone out as work is spent as mere feeling, which counts for nothing. It is only when the mind is very calm and collected that the whole of its energy is spent in doing good work. And if you read the lives of the great workers which the world has produced, you will find that they were wonderfully calm men. Nothing, as it were, could throw them off their balance. That is why the man who becomes angry never does a great amount of work, and the man whom nothing can make angry accomplishes so much. The man who gives way to anger, or hatred, or any other passion, cannot work; he only breaks himself to pieces, and does nothing practical. It is the calm, forgiving, equable, well-balanced mind that does the greatest amount of work.

The Vedanta preaches the ideal; and the ideal, as we know, is always far ahead of the real, of the practical, as we may call it. There are two tendencies in human nature: one to harmonise the ideal with the life, and the other to elevate the life to the ideal. It is a great thing to understand this, for the former tendency is the temptation of our lives. I think that I can only do a certain class of work. Most of it, perhaps, is bad; most of it, perhaps, has a motive power of passion behind it, anger, or greed, or selfishness. Now if any man comes to preach to me a certain ideal, the first step towards which is to give up selfishness, to give up self-enjoyment, I think that is impractical.

But when a man brings an ideal which can be reconciled with my selfishness, I am glad at once and jump at it. That is the ideal for me. As the word "orthodox" has been manipulated into various forms, so has been the word "practical". "My doxy is orthodoxy; your doxy is heterodoxy." So with practicality. What I think is practical, is to me the only practicality in the world. If I am a shopkeeper, I think shopkeeping the only practical pursuit in the world. If I am a thief, I think stealing is the best means of being practical; others are not practical. You see how we all use this word practical for things we like and can do.

Therefore I will ask you to understand that Vedanta, though it is intensely practical, is always so in the sense of the ideal. It does not preach an impossible ideal, however high it be, and it is high enough for an ideal. In one word, this ideal is that you are divine, "Thou art That". This is the essence of Vedanta; after all its ramifications and intellectual gymnastics, you know the human soul to be pure and omniscient, you see that such superstitions as birth and death would be entire nonsense when spoken of in connection with the soul. The soul was never born and will never die, and all these ideas that we are going to die and are afraid to die are mere superstitions. And all such ideas as that we can do this or cannot do that are superstitions. We can do everything.

The Vedanta teaches men to have faith in themselves first. As certain religions of the world say that a man who does not believe in a Personal God outside of himself is an atheist, so the Vedanta says, a man who does not believe in himself is an atheist. Not believing in the glory of our own soul is what the Vedanta calls atheism. Tomany this is, no doubt, a terrible idea; and most of us think that this ideal can never be reached; but the Vedanta insists that it can be realised by every one. There is neither man nor woman or child, nor difference of race or sex, nor anything that stands as a bar to the realisation of the ideal, because Vedanta shows that it is realised already, it is already there.

All the powers in the universe are already ours. It is we who have put our hands before our eyes and cry that it is dark. Know that there is no darkness around us. Take the hands away and there is the light which was from the beginning. Darkness never existed, weakness never existed. We who are fools cry that we are weak; we who are fools cry that we are impure. Thus Vedanta not only insists that the ideal is practical, but that it has been so all the time; and this Ideal, this Reality, is our own nature. Everything else that you see is false, untrue. As soon as you say, "I am a little mortal being," you are saying something which is not true, you are giving the lie to yourselves, you are hypnotising yourselves into something vile and weak and wretched.

The Vedanta recognises no sin, it only recognises error. And the greatest error, says the Vedanta, is to say that you are weak, that you are a sinner, a miserable creature, and that you have no power and you cannot do this and that. Every time you think in that way, you, as it were, rivet one more link in the chain that binds you down, you add one more layer of hypnotism on to your own soul. Therefore, whosoever thinks he is weak is wrong, whosoever thinks he is impure is wrong, and is throwing a bad thought into the world. This we must always bear in mind that in the Vedanta there is no attempt at reconciling the present life—the hypnotised life, this false life which we have assumed—with the ideal; but this false life must go, and the real life which is always existing must manifest itself, must shine out. No man becomes purer and purer, it is a matter of greater manifestation. The veil drops away, and the native purity of the soul begins to manifest itself. Everything is ours already—infinite purity, freedom, love, and power.

The Vedanta also says that not only can this be realised in the depths of forests or caves, but by men in all possible conditions of life. We have seen that the people who discovered these truths were neither living in caves nor forests, nor following the ordinary vocations of life, but men who, we have every reason to belive, led the busiest of lives, men who had to command armies, to sit on thrones, and look to the welfare of millions—and all these, in the days of absolute monarchy, and not as in these days when a king is to a great extent a mere figurehead. Yet they could find time to think out all these thoughts, to realise them, and to teach them to humanity.

How much more then should it be practical for us whose lives, compared with theirs, are lives of leisure? That we cannot realise them is a shame to us, seeing that we are comparatively free all the time, having very little to do. My requirements are as nothing compared with those of an ancient absolute monarch. My wants are as nothing compared with the demands of Arjuna on the battlefield of Kurukshetra, commanding a huge army; and yet he could find time in the midst of the din and turmoil of battle to talk the highest philosophy and to carry it into his life also. Surely we ought to be able to do as much in this life of ours—comparatively free, easy, and comfortable.

Most of us here have more time than we think we have, if we really want to use it for good. With the amount of freedom we have we can attain to two hundred ideals in this life, if we will, but we must not degrade the ideal to the actual. One of the most insinuating things comes to us in the shape of persons who apologise for our mistakes and teach us how to make special excuses for all our foolish wants and foolish desires; and we think that their ideal is the only ideal we need have. But it is not so. The Vedanta teaches no such things. The actual should be reconciled to the ideal, the present life should be made to coincide with life eternal.

For you must always remember that the one central ideal of Vedanta is this oneness. There are no two in anything, no two lives, nor even two different kinds of life for the two worlds. You will find the Vedas speaking of heavens and things like that at first; but later on, when they come to the highest ideals of their philosophy, they brush away all these things. There is but one life, one world, one existence. Everything is that—One, the difference is in degree and not in kind. The Vedanta entirely denies such ideas as that animals are separate from men, and that they were made and created by God to be used for our food. Some people have been kind enough to start an anti-vivisection society. I asked a member, "Why do you think, my friend, that it is quite lawful to kill animals for food, and not to kill one or two for scientific experiments?" He replied, "Vivisection is most horrible, but animals have been given to us for food." Oneness includes all animals. If man's life is immortal, so also is the animal's. The difference is only in degree and not in kind.

The amoeba and I are the same, the difference is only in degree; and from the standpoint of the highest life, all these differences vanish. A man may see a great deal of difference between grass and a little tree, but if you mount very high, the grass and the biggest tree will appear much the same. So, from the standpoint of the highest ideal, the lowest animal and the highest man are the same. If you believe there is a God, the animals and the highest creatures must be the same. A God who is partial to his children called men, and cruel to his children called brute beasts, is worse than a demon. I would rather die a hundred times than worship such a God. My whole life would be a fight with such a God.

But there is no difference, and those who say there is, are irresponsible, heartless people who do not know. Here is a case of the word practical used in a wrong sense. I myself may not be a very strict vegetarian, but I understand the ideal. When I eat meat I know it is wrong. Even if I am bound to eat it under certain circumstances, I know it is cruel. I must not drag my ideal down to the actual and apologize for my weak conduct in this way. The ideal is not to eat flesh, not to injure any being, for all animals are my brothers. If you can think of them as your brothers, you have made a little headway towards the brotherhood of all souls, not to speak of the brotherhood of man! That is child's play. You generally find that this is not very acceptable to many, because it teaches them to give up the actual, and go higher up to the ideal. But if you bring out a theory which is reconciled with their present conduct, they regard it as entirely practical.

There is this strongly conservative tendency in human nature: we do not like to move one step forward. I think of mankind just as I read of persons who become frozen in snow; all such, they say, want to go to sleep, and if you try to drag them up, they say, "Let me sleep; it is so beautiful to sleep in the snow", and they die there in that sleep. So is our nature. That is what we are doing all our life, getting frozen from the feet upwards, and yet wanting to sleep. Therefore you must struggle towards the ideal, and if a man comes who wants to bring that ideal down to your level, and teach a religion that does not carry that highest ideal, do not listen to him. To me that is an impracticable religion.

But if a man teaches a religion which presents the highest ideal, I am ready for him. Beware when anyone is trying to apologize for sense vanities and sense weaknesses. If anyone wants to preach that way to us, poor, sense-bound clods of earth as we have made ourselves by following that teaching, we shall never progress. I have seen many of these things, I have had some experience of the world, and my country is the land where religious sects grow like mushrooms. Every year new sects arise. But one thing I have marked, that it is only those that never want to reconcile the man of flesh with the man of truth that make progress. Wherever there is this false idea of reconciling fleshly vanities with the highest ideals, of dragging down God to the level of man, there comes decay. Man should not be degraded to worldly slavery, but should be raised up to God.

At the same time, there is another side to the question. We must not look down with contempt on others. All of us are going towards the same goal. The difference between weakness and strength is one of degree; the difference between virtue and vice is one of degree; the difference between heaven and hell is one of degree; the difference between life and death is one of degree; all differences in this world are of degree, and not of kind, because oneness is the secret of everything.

All is One, which manifests itself, either as thought, or life, or soul, or body, and the difference is only in degree. As such, we have no right to look down with contempt upon those who are not developed exactly in the same degree as we are. Condemn none; if you can stretch out a helping hand, do so. If you cannot, fold your hands, bless your brother, and let them go their own way. Dragging down and condemning is not the way to work. Never is work accomplished in that way. We spend our energies in condemning others. Criticism and condemnation is a vain way of spending our energies, for in the long run we come to learn that all are seeing the same thing, are more or less approaching the same ideal, and that most of our differences are merely differences of expression.

Take the idea of sin. I was telling you just now the Vedantic idea of it, and the other idea is that man is a sinner. They are practically the same, only the one takes the positive and the other the negative side. One shows to man his strength and the other his weakness. There may be weakness, says the Vedanta, but never mind, we want to grow. Disease was found out as soon as man was born. Everyone knows his disease; it requires no one to tell us what our diseases are.

But thinking all the time that we are diseased will not cure us—medicine is necessary. We may forget anything outside, we may try to become hypocrites to the external world, but in our heart of hearts we all know our weaknesses. But, says the Vedanta, being reminded of weakness does not help much; give strength, and strength does not come by thinking of weakness all the time. The remedy for weakness is not brooding over weakness, but thinking of strength. Teach men of the strength that is already within them. Instead of telling them they are sinners, the Vedanta takes the opposite position, and says, "You are pure and perfect, and what you call sin does not belong to you." Sins are very low degrees of Self-manifestation; manifest your Self in a high degree. That is the one thing to remember; all of us can do that. Never say, "No", never say, "I cannot", for you are infinite. Even time and space are as nothing compared with your nature. You can do anything and everything, you are almighty.

These are the principles of ethics, but we shall now come down lower and work out the details. We shall see how this Vedanta can be carried into our everyday life, the city life, the country life, the national life, and the home life of every nation. For, if a religion cannot help man wherever he may be, wherever he stands, it is not of much use; it will remain only a theory for the chosen few. Religion, to help mankind, must be ready and able to help him in whatever condition he is, in servitude or in freedom, in the depths of degradation or on the heights of purity; everywhere, equally, it should be able to come to his aid. The principle of Vedanta, or the ideal of religion, or whatever you may call it, will be fulfilled by its capacity for performing this great function.

The ideal of faith in ourselves is of the greatest help to us. If faith in ourselves had been more extensively taught and practised, I am sure a very large portion of the evils and miseries that we have would have vanished. Throughout the history of mankind, if any motive power has been more potent than another in the lives of all great men and women, it is that of faith in themselves. Born with the consciousness that they were to be great, they became great.

Let a man go down as low as possible; there must come a time when out of sheer desperation he will take an upward curve and will learn to have faith in himself. But it is better for us that we should know it from the very first. Why should we have all these bitter experiences in order to gain faith in ourselves? We can see that all the difference between man and man is owing to the existence of non-existence of faith in himself. Faith in ourselves will do everything. I have experienced it in my own life, and am still doing so; and as I grow older that faith is becoming stronger and stronger. He is an atheist who does not believe in himself.

The old religion said that he was an atheist who did not believe in God. The new religion says that he is the atheist who does not believe in himself. But it is not selfish faith, because the Vedanta, again, is the doctrine of oneness. It means faith in all, because you are all. Love for yourselves means love for all, love for animals, love for everything, for you are all one.

It is the great faith which will make the world better. I am sure of that. He is the highest man who can say with truth, "I know all about myself." Do you know how much energy, how many powers, how many forces are still lurking behind that frame of yours? What scientist has known all that is in man? Millions of years have passed since man first came here, and yet but one infinitesimal part of his powers has been manifested. Therefore, you must not say that you are weak. How do you know what possibilities lie behind that degradation on the surface? You know but little of that which is within you. For behind you is the ocean of infinite power and blessedness.

"This *Atman* is first to be heard of." Hear day and night that you are that Soul. Repeat it to yourselves day and night till it enters into your very veins, till it tingles in every drop of blood, till it is in your flesh and bone. Let the whole body be full of that one ideal, "I am the birthless, the deathless, the blissful, the omniscient, the omnipotent, ever-glorious Soul." Think on it day and night; think on it till it becomes part and parcel of your life. Meditate upon it, and out of that will come work. "Out of the fullness of the heart the mouth speaketh," and out of the fullness of the heart the hand worketh also.

Action will come. Fill yourselves with the idea; whatever you do, think well on it. All your actions will be magnified, transformed, deified, by the very power of the thought. If matter is powerful, thought is omnipotent. Bring this thought to bear upon your life, fill yourselves with the thought of your almightiness, your majesty, and your glory. Would to God no superstitions had been put into your head! Would to God we had not been surrounded from our birth by all these superstitious influences and paralysing ideas of our weakness and vileness! Would to God that mankind had had an easier path through which to attain to the noblest and highest truths! But man had to pass through all this; do not make the path more diffcult for those who are coming after you.

These are sometimes terrible doctrines to teach. I know people who get frightened at these ideas, but for those who want to be practical, this is the first thing to learn. Never tell yourselves or others that you are weak.

Do good if you can, but do not injure the world. You know in your inmost heart that many of your limited ideas, this humbling of yourself and praying and weeping to imaginary beings are superstitions. Tell me one case where these prayers have been answered. All the answers that came were from your own hearts. You know there are no ghosts, but no sooner are you in the dark than you feel a little creepy sensation. That is so because in our childhood we have had all these fearful ideas put into our heads. But do not teach these things to others through fear of society and public opinion, through fear of incurring the hatred of friends, or for fear of losing cherished superstitions. Be masters of all these.

What is there to be taught more in religion than the oneness of the universe and faith in one's self? All the works of mankind for thousands of years past have been towards this one goal, and mankind is yet working it out. It is your turn now and you already know the truth. For it has been taught on all sides. Not only philosophy and psychology, but materialistic sciences have declared it. Where is the scientific man today who fears to acknowledge the truth of this oneness of the universe? Who is there who dares talk of many worlds? All these are superstitions. There is only one life and one world, and this one life and one world is appearing to us as manifold. This manifoldness is like a dream. When you dream one dream passes away and another comes. You do not live in your dreams. The dreams come one after another, scene after scene unfolds before you. So it is in this world of ninety per cent misery and ten per cent happiness.

Perhaps after a while it will appear as ninety per cent happiness, and we shall call it heaven, but a time comes to the sage when the whole thing vanishes, and this world appears as God Himself, and his own soul as God. It is not therefore that there are many worlds, it is not that there are many lives. All this manifoldness is the manifestation of that One. That One is manifesting Himself as many, as matter, spirit, mind, thought, and everything else. It is that One, manifesting Himself as many. Therefore the first step for us to take is to teach the truth to ourselves and to others.

Let the world resound with this ideal, and let superstitions vanish. Tell it to men who are weak and persist in telling it. You are the Pure One; awake and arise, O mighty one, this sleep does not become you. Awake and arise, it does not befit you. Think not that you are weak and miserable. Almighty, arise and awake, and manifest your own nature. It is not fitting that you think yourself a sinner. It is not fitting that you think yourself weak. Say that to the world, say it to yourselves, and see what a practical result comes, see how with an electric flash everything is manifested, how everything is changed. Tell that to mankind, and show them their power. Then we shall learn how to apply it in our daily lives.

To be able to use what we call *Viveka* (discrimination), to learn how in every moment of our lives, in every one of our actions, to discriminate between what is right and wrong, true and false, we shall have to know the test of truth, which is purity, oneness. Everything that makes for oneness is truth. Love is truth, and hatred is false, because hatred makes for multiplicity. It is hatred that separates man from man; therefore it is wrong and false. It is a disintegrating power; it separates and destroys.

Love binds, love makes for that oneness. You become one, the mother with the child, families with the city, the whole world becomes one with the animals. For love is Existence, God Himself; and all this is the manifestation of that One Love, more or less expressed. The difference is only in degree, but it is the manifestation of that One Love throughout. Therefore in all our actions we have to judge whether it is making for diversity or for oneness. If for diversity we have to give it up, but if it makes for oneness we are sure it is good. So with our thoughts; we have to decide whether they make for disintegration, multiplicity, or for oneness, binding soul to soul and bringing one influence to bear. If they do this, we will take them up, and if not, we will throw them off as criminal.

The whole idea of ethics is that it does not depend on anything unknowable, it does not teach anything unknown, but in the language of the Upanishad, "The God whom you worship as an unknown God, the same I preach unto thee." It is through the Self that you know anything. I see the chair; but to see the chair, I have first to perceive myself and then the chair. It is in and through the Self that the chair is perceived. It is in and through the Self that you are known to me, that the whole world is known to me; and therefore to say this Self is unknown is sheer nonsense. Take off the Self and the whole universe vanishes.

In and through the Self all knowledge comes. Therefore it is the best known of all. It is yourself, that which you call I. You may wonder how this I of me can be the I of you. You may wonder how this limited I can be the unlimited Infinite, but it is so. The limited is a mere fiction. The Infinite has been covered up, as it were, and a little of It is manifesting as the I. Limitation can never come upon the unlimited; it is a fiction. The Self is known, therefore, to every one of us—man, woman, or child—and even to animals. Without knowing Him we can neither live nor move, nor have our being; without knowing this Lord of all, we cannot breathe or live a second. The God of the Vedanta is the most known of all and is not the outcome of imagination.

If this is not preaching a practical God, how else could you teach a practical God? Where is there a more practical God than He whom I see before me—a God omnipresent, in every being, more real than our senses? For you are He, the Omnipresent God Almighty, the Soul of your souls, and if I say you are not, I tell an untruth. I know it, whether at all times I realise it or not. He is the Oneness, the Unity of all, the Reality of all life and all existence.

These ideas of the ethics of Vedanta have to be worked out in detail, and, therefore, you must have patience. As I have told you, we want to take the subject in detail and work it up thoroughly, to see how the ideas grow from very low ideals, and how the one great ideal of oneness has developed and become shaped into the universal love; and we ought to study these in order to avoid dangers. The world cannot find time to work it up from the lowest steps. But what is the use of our standing on higher steps if we cannot give the truth to others coming afterwards?

Therefore, it is better to study it in all its workings; and first, it is absolutely necessary to clear the intellectual portion, although we know that intellectuality is almost nothing; for it is the heart that is of most importance. It is through the heart that the Lord is seen, and not through the intellect. The intellect is only the street-cleaner, cleansing the path for us, a secondary worker, the policeman; but the policeman is not a positive necessity for the workings of society. He is only to stop disturbances, to check wrong-doing, and that is all the work required of the intellect. When you read intellectual books, you think when you have mastered them, "Bless the Lord that I am out of them", because the intellect is blind and cannot move of itself, it has neither hands nor feet.

It is feeling that works, that moves with speed infinitely superior to that of electricity or anything else. Do you feel?—that is the question. If you do, you will see the Lord. It is the feeling that you have today that will be intensified, deified, raised to the highest platform, until it feels everything, the oneness in everything, till it feels God in itself and in others. The intellect can never do that. "Different methods of speaking words, different methods of explaining the texts of books, these are for the enjoyment of the learned, not for the salvation of the soul" (*Vivekachudamani*, 58).

Those of you who have read Thomas a Kempis know how in every page he insists on this, and almost every holy man in the world has insisted on it. Intellect is necessary, for without it we fall into crude errors and make all sorts of mistakes. Intellect checks these; but beyond that, do not try to build anything upon it. It is an inactive, secondary help; the real help is feeling, love. Do you feel for others? If you do, you are growing in oneness. If you do not feel for others, you may be the most intellectual giant ever born, but you will be nothing; you are but dry intellect, and you will remain so. And if you feel, even if you cannot read any book and do not know any language, you are in the right way. The Lord is yours.

Do you not know from the history of the world where the power of the prophets lay? Where was it? In the intellect? Did any of them write a fine book on philosophy, on the most intricate rationals of logic? Not one of them. They only spoke a few words. Feel like Christ and you will be a Christ; feel like Buddha and you will be a Buddha. It is feeling that is the life, the strength, the vitality, without which no amount of intellectual activity can reach God. Intellect is like limbs without the power of locomotion. It is only when feeling enters and gives them motion that they move and work on others. That is so all over the world, and it is a thing which you must always remember. It is one of the most practical things in Vedantic morality, for it is the teaching of the Vedanta that you are all Prophets, and all must be Prophets. The book is not the proof of your conduct, but you are the proof of the book. How do you know that a book teaches truth? Because you are truth and feel it. That is what the Vedanta says. What is the proof of the Christs and Buddhas of the world? That you and I feel like them.

That is how you and I understand that they were true. Our prophet-soul is the proof of their prophet-soul. Your godhead is the proof of God Himself. If you are not a prophet, there never has been anything true of God. If you are not God, there never was any God, and never will be. This, says the Vedanta, is the ideal to follow. Every one of us will have to become a prophet, and you are that already. Only know it. Never think there is anything impossible for the soul. It is the greatest heresy to think so. If there is sin, this is the only sin—to say that you are weak, or others are weak.

Practical Vedanta 2

I will tell to you a very ancient story from the Chhandogya Upanishad, which tells how knowledge came to a boy. The form of the story is very crude, but we shall find that it contains a principle. A young boy said to his mother, "I am going to study the Vedas. Tell me the name of my father and my caste." The mother was not a married woman, and in India the child of a woman who has not been married is considered an outcast; he is not recognised by society and is not entitled to study the Vedas.

So the poor mother said, "My child, I do not know your family name; I was in service, and served in different places; I do not know who your father is, but my name is Jabala and your name is Satyakama." The little child went to a sage and asked to be taken as a student. The sage asked him, "What is the name of your father, and what is your caste?" The boy repeated to him what he had heard from his mother. The sage at once said, "None but a Brahmin could speak such a damaging truth about himself. You are a Brahmin and I will teach you. You have not swerved from the truth." So he kept the boy with him and educated him.

This teacher gave Satyakama four hundred lean, weak cows to take care of, and sent him to the forest. There he went and lived for some time. The teacher had told him to come back when the herd would increase to the number of one thousand. After a few years, one day Satyakama heard a big bull in the herd saying to him, "We are a thousand now; take us back to your teacher. I will teach you a little of Brahman." "Say on, sir," said Satyakama. Then the bull said, "The East is a part of the Lord, so is the West, so is the South, so is the North.

The four cardinal points are the four parts of Brahman. Fire will also teach you something of Brahman." Fire was a great symbol in those days, and every student had to procure fire and make offerings. So on the following day, Satyakama started for his Guru's house, and when in the evening he had performed his oblation, and worshipped at the fire, and was sitting near it, he heard a voice come from the fire, "O Satyakama." "Speak, Lord," said Satyakama. (Perhaps you may remember a very similar story in the Old Testament, how Samuel heard a mysterious voice.) "O Satyakama, I am come to teach you a little of Brahman.

This earth is a portion of that Brahman. The sky and heaven are portions of it. The ocean is a part of that Brahman." Then the fire said that a certain bird would also teach him something. Satyakama continued his journey and on the next day when he had performed his evening sacrifice a swan came to him and said, "I will teach you something about Brahman. This fire which you worship, O Satyakama, is a part of that Brahman. The sun is a part, the moon is a part, the lightning is a part of that Brahman. A bird called Madgu will tell you more about it." The next evening that bird came, and a similar voice was heard by Satyakama, "I will tell you something about Brahman. Breath is a part of Brahman, sight is a part, hearing is a part, the mind is a part."

Then the boy arrived at his teacher's place and presented himself before him with due reverence. No sooner had the teacher seen this disciple than he remarked: "Satyakama, thy face shines like that of a knower of Brahman! Who then has taught thee?" "Beings other than men," replied Satyakama. "But I wish that you should teach me, sir. For I have heard from men like you that knowledge which is learnt from a Guru alone leads to the supreme good." Then the sage taught him the same knowledge which he had received from the gods. "And nothing was left out, yea, nothing was left out."

Now, apart from the allegories of what the bull, the fire, and the birds taught, we see the tendency of the thought and the direction in which it was going in those days. The great idea of which we here see the germ is that all these voices are inside ourselves. As we understand these truths better, we find that the voice is in our own heart, and the student understood that all the time he was hearing the truth; but his explanation was not correct.

He was interpreting the voice as coming from the external world, while all the time, it was within him. The second idea that we get is that of making the knowledge of the Brahman practical. The world is always seeking the practical possibilities of religion, and we find in these stories how it was becoming more and more practical everyday. The truth was shown through everything with which the students were familiar. The fire they were worshipping was Brahman, the earth was a part of Brahman, and so on.

The next story belongs to Upakoshala Kamalayana, a disciple of this Satyakama, who went to be taught by him and dwelt with him for some time. Now Satyakama went away on a journey, and the student became very down-hearted; and when the teacher's wife came and asked him why he was not eating, the boy said, "I am too unhappy to eat." Then a voice came from the fire he was worshipping, saying "This life is Brahman, Brahman is the ether, and Brahman is happiness. Know Brahman." "I know, sir," the boy replied, "that life is Brahman, but that It is ether and happiness I do not know."

Then it explained that the two words ether and happiness signified one thing in reality, *viz* the sentient (pure intelligence) that resides in the heart. So, it taught him Brahman as life and as the ether in the heart. Then the fire taught him, "This earth, food, fire, and sun whom you worship, are forms of Brahman. The person that is seen in the sun, I am He. He who knows this and meditates on Him, all his sins vanish and he has long life and becomes happy. He who lives in the cardinal points, the moon, the stars, and the water, I am He. He who lives in this life, the ether, the heavens, and the lightning, I am He." Here too we see the same idea of practical religion.

The things which they were worshipping, such as the fire, the sun, the moon, and so forth, and the voice with which they were familiar, form the subject of the stories which explain them and give them a higher meaning. And this is the real, practical side of Vedanta. It does not destroy the world, but it explains it; it does not destroy the person, but explains him; it does not destroy the individuality, but explains it by showing the real individuality. It does not show that this world is vain and does not exist, but it says, "Understand what this world is, so that it may not hurt you."

The voice did not say to Upakoshala that the fire which he was worshipping, or the sun, or the moon, or the lightning, or anything else, was all wrong, but it showed him that the same spirit which was inside the sun, and the moon, and lightning, and the fire, and the earth, was in him, so that everything became transformed, as it were, in the eyes of Upakoshala. The fire which was merely a material fire before, in which to make oblations, assumed a new aspect and became the Lord. The earth became transformed, life became transformed, the sun, the moon, the stars, the lightning, everything became transformed and deified. Their real nature was known.

The theme of the Vedanta is to see the Lord in everything, to see things in their real nature, not as they appear to be. Then another lesson is taught in the Upanishads: "He who shines through the eyes is Brahman; He is the Beautiful One, He is the Shining One. He shines in all these worlds." A certain peculiar light, a commentator says, which comes to the pure man, is what is meant by the light in the eyes, and it is said that when a man is pure such a light will shine in his eyes, and that light belongs really to the Soul within, which is everywhere. It is the same light which shines in the planets, in the stars, and suns.

I will now read to you some other doctrine of these ancient Upanishads, about birth and death and so on. Perhaps it will interest you. Shvetaketu went to the king of the Panchalas, and the king asked him, "Do you know where people go when they die? Do you know how they come back? Do you know why the other world does not become full?" The boy replied that he did not know. Then he went to his father and asked him the same questions. The father said, "I do not know," and he went to the king. The king said that this knowledge was never known to the priests, it was only with the kings, and that was the reason why kings ruled the world. This man stayed with the king for some time, for the king said he would teach him. "The other world, O Gautama, is the fire. The sun is its fuel. The rays are the smoke. The day is the flame.

The moon is the embers. And the stars are the sparks. In this fire the gods pour libation of faith and from this libation king Soma is born." So on he goes. "You need not make oblation to that little fire: the whole world is that fire, and this oblation, this worship, is continually going on. The gods, and the angels, and everybody is worshipping it. Man is the greatest symbol of fire, the body of man." Here also we see the ideal becoming practical and Brahman is seen in everything. The principle that underlies all these stories is that invented symbolism may be good and helpful, but already better symbols exist than any we can invent. You may invent an image through which to worship God, but a better image already exists, the living man. You may build a temple in which to worship God, and that may be good, but a better one, a much higher one, already exists, the human body.

You remember that the Vedas have two parts, the ceremonial and the knowledge portions. In time ceremonials had multiplied and become so intricate that it was almost hopeless to disentangle them, and so in the Upanishads we find that the ceremonials are almost done away with, but gently, by explaining them. We see that in old times they had these oblations and sacrifices, then the philosophers came, and instead of snatching away the symbols from the hands of the ignorant, instead of taking the negative position, which we unfortunately find so general in modern reforms, they gave them something to take their place. "Here is the symbol of fire," they said. "Very good! But here is another symbol, the earth. What a grand, great symbol! Here is this little temple, but the whole universe is a temple; a man can worship anywhere. There are the peculiar figures that men draw on the earth, and there are the altars, but here is the greatest of altars, the living, conscious human body, and to worship at this altar is far higher than the worship of any dead symbols."

We now come to a peculiar doctrine. I do not understand much of it myself. If you can make something out of it, I will read it to you. When a man dies, who has by meditation purified himself and got knowledge, he first goes to light, then from light to day, from day to the light-half of the moon, from that to the six months when the sun goes to the north, from that to the year, from the year to the sun, from the sun to the moon, from the moon to the lightning, and when he comes to the sphere of lightning, he meets a person who is not human, and that person leads him to (the conditioned) Brahman. This is the way of the gods. When sages and wise persons die, they go that way and they do not return.

What is meant by this month and year, and all these things, no one understands clearly. Each one gives his own meaning, and some say it is all nonsense. What is meant by going to the world of the moon and of the sun, and this person who comes to help the soul after it has reached the sphere of lightning, no one knows. There is an idea among the Hindus that the moon is a place where life exists, and we shall see how life has come from there. Those that have not attained to knowledge, but have done good work in this life, first go, when they die, through smoke, then to night, then to the dark fifteen days, then to the six months when the sun goes to the south, and from that they go to the region of their forefathers, then to ether, then to the region of the moon, and there become the food of the gods, and later, are born as gods and live there so long as their good works will permit.

And when the effect of the good work has been finished, they come back to earth by the same route. They first become ether, and then air, and then smoke, and then mist, then cloud, and then fall upon the earth as raindrops; then they get into food, which is eaten up by human beings, and finally become their children. Those whose works have been very good take birth in good families, and those whose works have been bad take bad births, even in animal bodies. Animals are continually coming to and going from this earth. That is why the earth is neither full nor empty.

Several ideas we can get also from this, and later on, perhaps, we shall be able to understand it better, and we can speculate a little upon what it means. The last part which deals with how those who have been in heaven return, is clearer, perhaps, than the first part; but the whole idea seems to be this that there is no permanent heaven without realising God. Now some people who have not realised God, but have done good work in this world, with the view of enjoying the results, go, when they die, through this and that place, until they reach heaven, and there they are born in the same way as we are here, as children of the gods, and they live there as long as their good works will permit. Out of this comes one basic idea of the Vedanta that everything which has name and form is transient.

This earth is transient, because it has name and form, and so the heavens must be transient, because there also name and form remain. A heaven which is eternal will be contradictory in terms, because everything that has name and form must begin in time, exist in time, and end in time. These are settled doctrines of the Vedanta, and as such the heavens are given up.

We have seen in the Samhita that the idea of heaven was that it was eternal, much the same as is prevalent among Mohammedans and Christians. The Mohammedans concretise it a little more. They say it is a place where there are gardens, beneath which rivers run. In the desert of Arabia water is very desirable, so the Mohammedan always conceives of his heaven as containing much water. I was born in a country where there are six months of rain every year. I should think of heaven, I suppose, as a dry place, and so also would the English people.

These heavens in the Samhita are eternal, and the departed have beautiful bodies and live with their forefathers, and are happy ever afterwards. There they meet with their parents, children, and other relatives, and lead very much the same sort of life as here, only much happier. All the difficulties and obstructions to happiness in this life have vanished, and only its good parts and enjoyments remain. But however comfortable mankind may consider this state of things, truth is one thing and comfort is another. There are cases where truth is not comfortable until we reach its climax. Human nature is very conservative. It does something, and having once done that, finds it hard to get out of it. The mind will not receive new thoughts, because they bring discomfort.

In the Upanishads, we see a tremendous departure made. It is declared that these heavens in which men live with the ancestors after death cannot be permanent, seeing that everything which has name and form must die.

If there are heavens with forms, these heavens must vanish in course of time; they may last millions of years, but there must come a time when they will have to go. With this idea came another that these souls must come back to earth, and that heavens are places where they enjoy the results of their good works, and after these effects are finished they come back into this earth life again. One thing is clear from this that mankind had a perception of the philosophy of causation even at the early time. Later on we shall see how our philosophers bring that out in the language of philosophy and logic, but here it is almost in the language of children. One thing you may remark in reading these books that it is all internal perception.

If you ask me if this can be practical, my answer is, it has been practical first, and philosophical next. You can see that first these things have been perceived and realised and then written. This world spoke to the early thinkers. Birds spoke to them, animals spoke to them, the sun and the moon spoke to them; and little by little they realised things, and got into the heart of nature. Not by cognition, not by the force of logic, not by picking the brains of others and making a big book, as is the fashion in modern times, not even as I do, by taking up one of their writings and making a long lecture, but by patient investigation and discovery they found out the truth.

Its essential method was practice, and so it must be always. Religion is ever a practical science, and there never was nor will be any theological religion. It is practice first, and knowledge afterwards. The idea that souls come back is already there. Those persons who do good work with the idea of a result, get it, but the result is not permanent. There we get the idea of causation very beautifully put forward, that the effect is only commensurate with the cause. As the cause is, so the effect will be. The cause being finite, the effect must be finite. If the cause is eternal the effect can be eternal, but all these causes, doing good work, and all other things, are only finite causes, and as such cannot produce infinite result.

We now come to the other side of the question. As there cannot be an eternal heaven, on the same grounds, there cannot be an eternal hell. Suppose I am a very wicked man, doing evil every minute of my life. Still, my whole life here, compared with my eternal life, is nothing. If there be an eternal punishment, it will mean that there is an infinite effect produced by a finite cause, which cannot be. If I do good all my life, I cannot have an infinite heaven; it would be making the same mistake. But there is a third course which applies to those who have known the Truth, to those who have realised it. This is the only way to get beyond this veil of Maya—to realise what Truth is; and the Upanishads indicate what is meant by realising the Truth.

It means recognising neither good nor bad, but knowing all as coming from the Self; Self is in everything. It means denying the universe; shutting your eyes to it; seeing the Lord in hell as well as in heaven; seeing the Lord in death as well as in life. This is the line of thought in the passage I have read to you; the earth is a symbol of the Lord, the sky is the Lord, the place we fill is the Lord, everything is Brahman. And this is to be seen, realised, not simply talked or thought about. We can see as its logical consequence that when the soul has realised that everything is full of the Lord, of Brahman, it will not care whether it goes to heaven, or hell, or anywhere else; whether it be born again on this earth or in heaven. These things have ceased to have any meaning to that soul, because every place is the same, every place is the temple of the Lord, every place has become holy and the presence of the Lord is all that it sees in heaven, or hell, or anywhere else. Neither good nor bad, neither life nor death—only the one infinite Brahman exists.

According to the Vedanta, when a man has arrived at that perception, he has become free, and he is the only man who is fit to live in this world. Others are not. The man who sees evil, how can he live in this world? His life is a mass of misery. The man who sees dangers, his life is a misery; the man who sees death, his life is a misery. That man alone can live in this world, he alone can say, "I enjoy this life, and I am happy in this life," who has seen the Truth, and the Truth in everything. By the by, I may tell you that the idea of hell does not occur in the Vedas anywhere. It comes with the Puranas much later. The worst punishment according to the Vedas is coming back to earth, having another chance in this world. From the very first we see the idea is taking the impersonal turn. The ideas of punishment and reward are very material, and they are only consonant with the idea of a human God, who loves one and hates another, just as we do. Punishment and reward are only admissible with the existence of such a God.

They had such a God in the Samhita, and there we find the idea of fear entering, but as soon as we come to the Upanishads, the idea of fear vanishes, and the impersonal idea takes its place. It is naturally the hardest thing for man to understand, this impersonal idea, for he is always clinging on to the person. Even people who are thought to be great thinkers get disgusted at the idea of the Impersonal God. But to me it seems so absurd to think of God as an embodied man. Which is the higher idea, a living God, or a dead God? A God whom nobody sees, nobody knows, or a God Known?

The Impersonal God is a living God, a principle. The difference between personal and impersonal is this, that the personal is only a man, and the impersonal idea is that He is the angel, the man, the animal, and yet something more which we cannot see, because impersonality includes all personalities, is the sum total of everything in the universe, and infinitely more besides. "As the one fire coming into the world is manifesting itself in so many forms, and yet is infinitely more besides," so is the Impersonal.

We want to worship a living God. I have seen nothing but God all my life, nor have you. To see this chair you first see God, and then the chair in and through Him. He is everywhere saying, "I am". The moment you feel "I am", you are conscious of Existence. Where shall we go to find God if we cannot see Him in our own hearts and in every living being? "Thou art the man, Thou art the woman, Thou art the girl, and Thou art the boy. Thou art the old man tottering with a stick.

Thou art the young man walking in the pride of his strength." Thou art all that exists, a wonderful living God who is the only fact in the universe. This seems to many to be a terrible contradiction to the traditional God who lives behind a veil somewhere and whom nobody ever sees. The priests only give us an assurance that if we follow them, listen to their admonitions, and walk in the way they mark out for us—then when we die, they will give us a passport to enable us to see the face of God! What are all these heaven ideas but simply modifications of this nonsensical priestcraft?

Of course the impersonal idea is very destructive; it takes away all trade from the priests, churches, and temples. In India there is a famine now, but there are temples in each one of which there are jewels worth a king's ransom! If the priests taught this Impersonal idea to the people, their occupation would be gone. Yet we have to teach it unselfishly, without priestcraft. You are God and so am I; who obeys whom? Who worships whom? You are the highest temple of God; I would rather worship you than any temple, image, or Bible. Why are some people so contradictory in their thought? They are like fish slipping through our fingers.

They say they are hard-headed practical men. Very good. But what is more practical than worshipping here, worshipping you? I see you, feel you, and I know you are God. The Mohammedan says, there is no God but Allah. The Vedanta says, there is nothing that is not God. It may frighten many of you, but you will understand it by degrees.

The living God is within you, and yet you are building churches and temples and believing all sorts of imaginary nonsense. The only God to worship is the human soul in the human body. Of course all animals are temples too, but man is the highest, the Taj Mahal of temples. If I cannot worship in that, no other temple will be of any advantage. The moment I have realised God sitting in the temple of every human body, the moment I stand in reverence before every human being and see God in him—that moment I am free from bondage, everything that binds vanishes, and I am free.

This is the most practical of all worship. It has nothing to do with theorising and speculation. Yet it frightens many. They say it is not right. They go on theorising about old ideals told them by their grandfathers, that a God somewhere in heaven had told some one that he was God. Since that time we have only theories. This is practicality according to them, and our ideas are impractical! No doubt, the Vedanta says that each one must have his own path, but the path is not the goal.

The worship of a God in heaven and all these things are not bad, but they are only steps towards the Truth and not the Truth itself. They are good and beautiful, and some wonderful ideas are there, but the Vedanta says at every point, "My friend, Him whom you are worshipping as unknown, I worship as thee. He whom you are worshipping as unknown and are seeking for, throughout the universe, has been with you all the time. You are living through Him, and He is the Eternal Witness of the universe." "He whom all the Vedas worship, nay, more, He who is always present in the eternal 'I', He is existing, the whole universe exists. He is the light and life of the universe.

If the 'I' were not in you, you would not see the sun, everything would be a dark mass. He shining, you see the world."

One question is generally asked, and it is this that this may lead to a tremendous amount of difficulty. Everyone of us will think, "I am God, and whatever I do or think must be good, for God can do no evil." In the first place, even taking this danger of misinterpretation for granted, can it be proved that on the other side the same danger does not exist? They have been worshipping a God in heaven separate from them, and of whom they are much afraid. They have been born shaking with fear, and all their life they will go on shaking. Has the world been made much better by this?

Those who have understood and worshipped a Personal God, and those who have understood and worshipped an Impersonal God, on which side have been the great workers of the world—gigantic workers, gigantic moral powers? Certainly on the Impersonal. How can you expect morality to be developed through fear? It can never be. "Where one sees another, where one hears another, that is Maya. When one does not see another, when one does not hear another, when everything has become the *Atman*, who sees whom, who perceives whom?" It is all He, and all I, at the same time. The soul has become pure. Then, and then alone we understand what love is. Love cannot come through fear, its basis is freedom. When we really begin to love the world, then we understand what is meant by brotherhood or mankind, and not before.

So, it is not right to say that the Impersonal idea will lead to a tremendous amount of evil in the world, as if the other doctrine never lent itself to works of evil, as if it did not lead to sectarianism deluging the world with blood and causing men to tear each other to pieces. "My God is the greatest god, let us decide it by a free fight." That is the outcome of dualism all over the world. Come out into the broad open light of day, come out from the little narrow paths, for how can the infinite soul rest content to live and die in small ruts? Come out into the universe of Light. Everything in the universe is yours, stretch out your arms and embrace it with love. If you ever felt you wanted to do that, you have felt God.

You remember that passage in the sermon of Buddha, how he sent a thought of love towards the south, the north, the east, and the west, above and below, until the whole universe was filled with this love, so grand, great, and infinite. When you have that feeling, you have true personality. The whole universe is one person; let go the little things. Give up the small for the Infinite, give up small enjoyments for infinite bliss.

It is all yours, for the Impersonal includes the Personal. So God is Personal and Impersonal at the same time. And Man, the Infinite, Impersonal Man, is manifesting Himself as person. We the infinite have limited ourselves, as it were, into small parts. The Vedanta says that Infinity is our true nature; it will never vanish, it will abide for ever. But we are limiting ourselves by our Karma, which like a chain round our necks has dragged us into this limitation. Break that chain and be free. Trample law under your feet. There is no law in human nature, there is no destiny, no fate.

How can there be law in infinity? Freedom is its watchword. Freedom is its nature, its birth-right. Be free, and then have any number of personalities you like. Then we will play like the actor who comes upon the stage and plays the part of a beggar. Contrast him with the actual beggar walking in the streets. The scene is, perhaps, the same in both cases, the words are, perhaps, the same, but yet what difference! The one enjoys his beggary while the other is suffering misery from it. And what makes this difference? the one is free and the other is bound. The actor knows his beggary is not true, but that he has assumed it for play, while the real beggar thinks that it is his too familiar state and that he has to bear it whether he wills it or not.

This is the law. So long as we have no knowledge of our real nature, we are beggars, jostled about by every force in nature; and made slaves of by everything in nature; we cry all over the world for help, but help never comes to us; we cry to imaginary beings, and yet it never comes. But still we hope help will come, and thus in weeping, wailing, and hoping, one life is passed, and the same play goes on and on.

Be free; hope for nothing from anyone. I am sure if you look back upon your lives you will find that you were always vainly trying to get help from others which never came. All the help that has come was from within yourselves. You only had the fruits of what you yourselves worked for, and yet you were strangely hoping all the time for help. A rich man's parlour is always full; but if you notice, you do not find the same people there. The visitors are always hoping that they will get something from those wealthy men, but they never do.

So are our lives spent in hoping, hoping, hoping, which never comes to an end. Give up hope, says the Vedanta. Why should you hope? You have everything, nay, you are everything. What are you hoping for? If a king goes mad, and runs about trying to find the king of his country, he will never find him, because he is the king himself. He may go through every village and city in his own country, seeking in every house, weeping and wailing, but he will never find him, because he is the king himself. It is better that we know we are God and give up this fool's search after Him; and knowing that we are God we become happy and contented. Give up all these mad pursuits, and then play your part in the universe, as an actor on the stage.

The whole vision is changed, and instead of an eternal prison this world has become a playground; instead of a land of competition it is a land of bliss, where there is perpetual spring, flowers bloom and butterflies flit about.

This very world becomes heaven, which formerly was hell. To the eyes of the bound it is a tremendous place of torment, but to the eyes of the free it is quite otherwise. This one life is the universal life, heavens and all those places are here. All the gods are here, the prototypes of man. The gods did not create man after their type, but man created gods. And here are the prototypes, here is Indra, here is Varuna, and all the gods of the universe. We have been projecting our little doubles, and we are the originals of these gods, we are the real, the only gods to be worshipped.

This is the view of the Vedanta, and this its practicality. When we have become free, we need not go mad and throw up society and rush off to die in the forest or the cave; we shall remain where we were, only we shall understand the whole thing. The same phenomena will remain, but with a new meaning. We do not know the world yet; it is only through freedom that we see what it is, and understand its nature. We shall see then that this so-called law, or fate, or destiny occupies only an infinitesimal part of our nature.

It was only one side, but on the other side there was freedom all the time. We did not know this, and that is why we have been trying to save ourselves from evil by hiding our faces in the ground, like the hunted hare. Through delusion we have been trying to forget our nature, and yet we could not; it was always calling upon us, and all our search after God or gods, or external freedom, was a search after our real nature. We mistook the voice. We thought it was from the fire, or from a god or the sun, or moon, or stars, but at last we have found that it was from within ourselves. Within ourselves is this eternal voice speaking of eternal freedom; its music is eternally going on. Part of this music of the Soul has become the earth, the law, this universe, but it was always ours and always will be. In one word, the ideal of Vedanta is to know man as he really is, and this is its message, that if you cannot worship your brother man, the manifested God, how can you worship a God who is unmanifested?

Do you not remember what the Bible says, "If you cannot love your brother whom you have seen, how can you love God whom you have not seen?" If you cannot see God in the human face, how can you see him in the clouds, or in images made of dull, dead matter, or in mere fictitious stories of our brain? I shall call you religious from the day you begin to see God in men and women, and then you will understand what is meant by turning the left cheek to the man who strikes you on the right. When you see man as God, everything, even the tiger, will be welcome. Whatever comes to you is but the Lord, the Eternal, the Blessed One, appearing to us in various forms, as our father, and mother, and friend, and child—they are our own soul playing with us.

As our human relationships can thus be made divine, so our relationship with God may take any of these forms and we can look upon Him as our father, or mother, or friend, or beloved. Calling God Mother is a higher ideal than calling Him Father; and to call Him Friend is still higher; but the highest is to regard Him as the Beloved. The highest point of all is to see no difference between lover and beloved. You may remember, perhaps, the old Persian story, of how a lover came and knocked at the door of the beloved and was asked, "Who are you?" He answered, "It is I", and there was no response. A second time he came, and exclaimed, "I am here", but the door was not opened. The third time he came, and the voice asked from inside, "Who is there?" He replied, "I am thyself, my beloved", and the door opened. So is the relation between God and ourselves. He is in everything. He is everything. Every man and woman is the palpable, blissful, living God. Who says God is unknown? Who says He is to be searched after?

We have found God eternally. We have been living in Him eternally; everywhere He is eternally known, eternally worshipped.

Then comes another idea, that other forms of worship are not errors. This is one of the great points to be remembered, that those who worship God through ceremonials and forms, however crude we may think them to be, are not in error. It is the journey from truth to truth, from lower truth to higher truth. Darkness is less light; evil is less good; impurity is less purity. It must always be borne in mind that we should see others with eyes of love, with sympathy, knowing that they are going along the same path that we have trodden.

If you are free, you must know that all will be so sooner or later, and if you are free, how can you see the impermanent? If you are really pure, how do you see the impure? For what is within, is without. We cannot see impurity without having it inside ourselves. This is one of the practical sides of Vedanta, and I hope that we shall all try to carry it into our lives. Our whole life here is to carry this into practice, but the one great point we gain is that we shall work with satisfaction and contentment, instead of with discontent and dissatisfaction, for we know that Truth is within us, we have it as our birth-right, and we have only to manifest it, and make it tangible.

Practical Vedanta 3

In the Chhandogya Upanishad we read that a sage called Narada came to another called Sanatkumara, and asked him various questions, of which one was, if religion was the cause of things as they are. And Sanatkumara leads him, as it were, step by step, telling him that there is something higher than this earth, and something higher than that, and so on, till he comes to *Akasha*. Ether is higher than light, because in the ether are the sun and the moon, lightning and the stars; in ether we live, and in ether we die. Then the question arises, if there is anything higher than that, and Sanatkumara tells him of Prana. This Prana, according to the Vedanta, is the principle of life. It is like ether, an omnipresent principle; and all motion, either in the body or anywhere else, is the work of this Prana. It is greater than *Akasha*, and through it everything lives. Prana is in the mother, in the father, in the sister, in the teacher, Prana is the knower.

I will read another passage, where Shvetaketu asks his father about the Truth, and the father teaches him different things, and concludes by saying, "That which is the fine cause in all these things, of it are all these things made. That is the All, that is Truth, thou art That, O Shvetaketu." And then he gives various examples. "As a bee, O Shvetaketu, gathers honey from different flowers, and as the different honeys do not know that they are from various trees, and from various flowers, so all of us, having come to that Existence, know not that we have done so. Now, that which is that subtle essence, in it all that exists has its self. It is the True. It is the Self and thou, O Shvetaketu, are That." He gives another example of the rivers running down to the ocean. "As the rivers, when they are in the ocean, do not know that they have been various rivers, even so when we come out of that Existence, we do not know that we are That. O Shvetaketu, thou are That." So on he goes with his teachings.

Now there are two principles of knowledge. The one principle is that we know by referring the particular to the general, and the general to the universal; and the second is that anything of which the explanation is sought is to be explained so far as possible from its own nature. Taking up the first principle, we see that all our knowledge really consists of classifications, going higher and higher. When something happens singly, we are, as it were, dissatisfied. When it can be shown that the same thing happens again and again, we are satisfied and call it law. When we find that one apple falls, we are dissatisfied; but when we find that all apples fall, we call it the law of gravitation and are satisfied. The fact is that from the particular we deduce the general.

When we want to study religion, we should apply this scientific process. The same principle also holds good here, and as a fact we find that that has been the method all through. In reading these books from which I have been translating to you, the earliest idea that I can trace is this principle of going from the particular to the general. We see how the "bright ones" became merged into one principle; and likewise in the ideas of the cosmos we find the ancient thinkers going higher and higher—from the fine elements they go to finer and more embracing elements, and from these particulars they come to one omnipresent and from that even they go to an all-embracing force, or Prana; and through all this runs the principle, that one is not separate from the others. It is the very ether that exists in the higher form of Prana, or the higher form of Prana concretes, so to say, and becomes ether; and that becomes still grosser, and so on. The generalisation of the Personal God is another case in point. We have seen how this generalisation was reached, and was called the sum total of all consciousness. But a difficulty arises—it is an incomplete generalisation. We take up only one side of the facts of nature, the fact of consciousness, and upon that we generalise, but the other side is left out.

So, in the first place it is a defective generalisation. There is another insufficiency, and that relates to the second principle. Everything should be explained from its own nature. There may have been people who thought that every apple that fell to the ground was dragged down by a ghost, but the explanation is the law of gravitation; and although we know it is not a perfect explanation, yet it is much better than the other, because it is derived from the nature of the thing itself, while the other posits an extraneous cause. So throughout the whole range of our knowledge; the explanation which is based upon the nature of the thing itself is a scientific explanation, and an explanation which brings in an outside agent is unscientific.

So the explanation of a Personal God as the creator of the universe has to stand that test. If that God is outside of nature, having nothing to do with nature, and this nature is the outcome of the command of that God and produced from nothing, it is a very unscientific theory, and this has been the weak point of every Theistic religion throughout the ages. These two defects we find in what is generally called the theory of monotheism, the theory of a Personal God, with all the qualities of a human being multiplied very much, who, by His will, created this universe out of nothing and yet is separate from it. This leads us into two difficulties.

As we have seen, it is not a sufficient generalisation, and secondly, it is not an explanation of nature from nature. It holds that the effect is not the cause, that the cause is entirely separate from the effect. Yet all human knowledge shows that the effect is but the cause in another form. To this idea the discoveries of modern science are tending every day, and the latest theory that has been accepted on all sides is the theory of evolution, the principle of which is that the effect is but the cause in another form, a readjustment of the cause, and the cause takes the form of the effect. The theory of creation out of nothing would be laughed at by modern scientists.

Now, can religion stand these tests? If there be any religious theories which can stand these two tests, they will be acceptable to the modern mind, to the thinking mind. Any other theory which we ask the modern man to believe, on the authority of priests, or churches, or books, he is unable to accept, and the result is a hideous mass of unbelief. Even in those in whom there is an external display of belief, in their hearts there is a tremendous amount of unbelief. The rest shrink away from religion, as it were, give it up, regarding it as priestcraft only.

Religion has been reduced to a sort of national form. It is one of our very best social remnants; let it remain. But the real necessity which the grandfather of the modern man felt for it is gone; he no longer finds it satisfactory to his reason. The idea of such a Personal God, and such a creation, the idea which is generally known as monotheism in every religion, cannot hold its own any longer. In India it could not hold its own because of the Buddhists, and that was the very point where they gained their victory in ancient times. They showed that if we allow that nature is possessed of infinite power, and that nature can work out all its wants, it is simply unnecessary to insist that there is something besides nature. Even the soul is unnecessary.

The discussion about substance and qualities is very old, and you will sometimes find that the old superstition lives even at the present day. Most of you have read how, during the Middle Ages, and, I am sorry to say, even much later, this was one of the subjects of discussion, whether qualities adhered to substance, whether length, breadth, and thickness adhered to the substance which we call dead matter, whether, the substance remaining, the qualities are there or not. To this our Buddhist says, "You have no ground for maintaining the existence of such a substance; the qualities are all that exist; you do not see beyond them." This is just the position of most of our modern agnostics.

For it is this fight of the substance and qualities that, on a higher plane, takes the form of the fight between noumenon and phenomenon. There is the phenomenal world, the universe of continuous change, and there is something behind which does not change; and this duality of existence, noumenon and phenomenon, some hold, is true, and others with better reason claim that you have no right to admit the two, for what we see, feel, and think is only the phenomenon. You have no right to assert there is anything beyond phenomenon; and there is no answer to this.

The only answer we get is from the monistic theory of the Vedanta. It is true that only one exists, and that one is either phenomenon or noumenon. It is not true that there are two—something changing, and, in and through that, something which does not change; but it is the one and the same thing which appears as changing, and which is in reality unchangeable. We have come to think of the body, and mind, and soul as many, but really there is only one; and that one is appearing in all these various forms.

Take the well-known illustration of the monists, the rope appearing as the snake. Some people, in the dark or through some other cause, mistake the rope for the snake, but when knowledge comes, the snake vanishes and it is found to be a rope. By this illustration we see that when the snake exists in the mind, the rope has vanished, and when the rope exists, the snake has gone.

When we see phenomenon, and phenomenon only, around us, the noumenon has vanished, but when we see the noumenon, the unchangeable, it naturally follows that the phenomenon has vanished. Now, we understand better the position of both the realist and the idealist. The realist sees the phenomenon only, and the idealist looks to the noumenon. For the idealist, the really genuine idealist, who has truly arrived at the power of perception, whereby he can get away from all ideas of change, for him the changeful universe has vanished, and he has the right to say it is all delusion, there is no change. The realist at the same time looks at the changeful. For him the unchangeable has vanished, and he has a right to say this is all real. What is the outcome of this philosophy? It is that the idea of Personal God is not sufficient. We have to get to something higher, to the Impersonal idea. It is the only logical step that we can take. Not that the personal idea would be destroyed by that, not that we supply proof that the Personal God does not exist, but we must go to the Impersonal for the explanation of the personal, for the Impersonal is a much higher generalisation than the personal. The Impersonal only can be Infinite, the personal is limited.

Thus we preserve the personal and do not destroy it. Often the doubt comes to us that if we arrive at the idea of the Impersonal God, the personal will be destroyed, if we arrive at the idea of the Impersonal man, the personal will be lost. But the Vedantic idea is not the destruction of the individual, but its real preservation. We cannot prove the individual by any other means but by referring to the universal, by proving that this individual is really the universal. If we think of the individual as separate from everything else in the universe, it cannot stand a minute. Such a thing never existed.

Secondly, by the application of the second principle, that the explanation of everything must come out of the nature of the thing, we are led to a still bolder idea, and one more difficult to understand. It is nothing less than this, that the Impersonal Being, our highest generalisation, is in ourselves, and we are That. "O Shvetaketu, thou art That." You are that Impersonal Being; that God for whom you have been searching all over the universe is all the time yourself—yourself not in the personal sense but in the Impersonal. The man we know now, the manifested, is personalised, but the reality of this is the Impersonal. To understand the personal we have to refer it to the Impersonal, the particular must be referred to the general, and that Impersonal is the Truth, the Self of man.

There will be various questions in connection with this, and I shall try to answer them as we go on. Many difficulties will arise, but first let us clearly understand the position of monism. As manifested beings we appear to be separate, but our reality is one, and the less we think of ourselves as separate from that One, the better for us. The more we think of ourselves as separate from the Whole, the more miserable we become. From this monistic principle we get at the basis of ethics, and I venture to say that we cannot get any ethics from anywhere else. We know that the oldest idea of ethics was the will of some particular being or beings, but few are ready to accept that now, because it would be only a partial generalisation.

The Hindus say we must not do this or that because the Vedas say so, but the Christian is not going to obey the authority of the Vedas. The Christian says you must do this and to do that because the Bible says so. That will not be binding on those who do not believe in the Bible. But we must have a theory which is large enough to take in all these various grounds. Just as there are millions of people who are ready to believe in a Personal Creator, there have also been thousands of the brightest minds in this world who felt that such ideas were not sufficient for them, and wanted something higher, and wherever religion was not broad enough to include all these minds, the result was that the brightest minds in society were always outside of religion; and never was this so marked as at the present time, especially in Europe.

To include these minds, therefore, religion must become broad enough. Everything it claims must be judged from the standpoint of reason. Why religions should claim that they are not bound to abide by the standpoint of reason, no one knows. If one does not take the standard of reason, there cannot be any true judgment, even in the case of religions. One religion may ordain something very hideous. For instance, the Mohammedan religion allows Mohammedans to kill all who are not of their religion.

It is clearly stated in the Koran, "Kill the infidels if they do not become Mohammedans." They must be put to fire and sword. Now if we tell a Mohammedan that this is wrong, he will naturally ask, "How do you know that? How do you know it is not good? My book says it is." If you say your book is older, there will come the Buddhist, and say, my book is much older still. Then will come the Hindu, and say, my books are the oldest of all. Therefore referring to books will not do. Where is the standard by which you can compare? You will say, look at the Sermon on the Mount, and the Mohammedan will reply, look at the Ethics of the Koran.

The Mohammedan will say, who is the arbiter as to which is the better of the two? Neither the New Testament nor the Koran can be the arbiter in a quarrel between them. There must be some independent authority, and that cannot be any book, but something which is universal; and what is more universal than reason? It has been said that reason is not strong enough; it does not always help us to get at the Truth; many times it makes mistakes, and, therefore, the conclusion is that we must believe in the authority of a church! That was said to me by a Roman Catholic, but I could not see the logic of it. On the other hand I should say, if reason be so weak, a body of priests would be weaker, and I am not going to accept their verdict, but I will abide by my reason, because with all its weakness there is some chance of my getting at truth through it; while, by the other means, there is no such hope at all.

We should, therefore, follow reason and also sympathise with those who do not come to any sort of belief, following reason. For it is better that mankind should become atheist by following reason than blindly believe in two hundred millions of gods on the authority of anybody. What we want is progress, development, realisation. No theories ever made men higher. No amount of books can help us to become purer. The only power is in realisation, and that lies in ourselves and comes from thinking. Let men think. A clod of earth never thinks; but it remains only a lump of earth. The glory of man is that he is a thinking being. It is the nature of man to think and therein he differs from animals. I believe in reason and follow reason having seen enough of the evils of authority, for I was born in a country where they have gone to the extreme of authority.

The Hindus believe that creation has come out of the Vedas. How do you know there is a cow? Because the word cow is in the Vedas. How do you know there is a man outside? Because the word man is there. If it had not been, there would have been no man outside. That is what they say. Authority with a vengeance! And it is not studied as I have studied it, but some of the most powerful minds have taken it up and spun out wonderful logical theories round it. They have reasoned it out, and there it stands—a whole system of philosophy; and thousands of the brightest intellects have been dedicated through thousands of years to the working out of this theory. Such has been the power of authority, and great are the dangers thereof. It stunts the growth of humanity, and we must not forget that we want growth. Even in all relative truth, more than the truth itself, we want the exercise. That is our life.

The monistic theory has this merit that it is the most rational of all the religious theories that we can conceive of. Every other theory, every conception of God which is partial and little and personal is not rational. And yet monism has this grandeur that it embraces all these partial conceptions of God as being necessary for many. Some people say that this personal explanation is irrational. But it is consoling; they want a consoling religion and we understand that it is necessary for them. The clear light of truth very few in this life can bear, much less live up to.

It is necessary, therefore, that this comfortable religion should exist; it helps many souls to a better one. Small minds whose circumference is very limited and which require little things to build them up, never venture to soar high in thought. Their conceptions are very good and helpful to them, even if only of little gods and symbols. But you have to understand the Impersonal, for it is in and through that alone that these others can be explained. Take, for instance, the idea of a Personal God. A man who understands and believes in the Impersonal—John Stuart Mill, for example—may say that a Personal God is impossible, and cannot be proved. I admit with him that a Personal God cannot be demonstrated.

But He is the highest reading of the Impersonal that can be reached by the human intellect, and what else is the universe but various readings of the Absolute? It is like a book before us, and each one has brought his intellect to read it, and each one has to read it for himself. There is something which is common in the intellect of all men; therefore certain things appear to be the same to the intellect of mankind. That you and I see a chair proves that there is something common to both our minds. Suppose a being comes with another sense, he will not see the chair at all; but all beings similarly constituted will see the same things.

Thus this universe itself is the absolute, the unchangeable, the noumenon; and the phenomenon constitutes the reading thereof. For you will first find that all phenomena are finite. Every phenomenon that we can see, feel, or think of, is finite, limited by our knowledge, and the Personal God as we conceive of Him is in fact a phenomenon. The very idea of causation exists only in the phenomenal world, and God as the cause of this universe must naturally be thought of as limited, and yet He is the same Impersonal God. This very universe, as we have seen, is the same Impersonal Being read by our intellect. Whatever is reality in the universe is that Impersonal Being, and the forms and conceptions are given to it by our intellects. Whatever is real in this table is that Being, and the table form and all other forms are given by our intellects.

Now, motion, for instance, which is a necessary adjunct of the phenomenal, cannot be predicated of the Universal. Every little bit, every atom inside the universe, is in a constant state of change and motion, but the universe as a whole is unchangeable, because motion or change is a relative thing; we can only think of something in motion in comparison with something which is not moving. There must be two things in order to understand motion. The whole mass of the universe, taken as a unit, cannot move. In regard to what will it move?

It cannot be said to change. With regard to what will it change? So the whole is the absolute; but within it every particle is in a constant state of flux and change. It is unchangeable and changeable at the same time, Impersonal and Personal in one. This is our conception of the universe, of motion and of God, and that is what is meant by "Thou art That". Thus we see that the Impersonal instead of dong away with the personal, the Absolute instead of pulling down the relative, only explains it to the full satisfaction of our reason and heart. The Personal God and all that exists in the universe are the same Impersonal Being seen through our minds. When we shall be rid of our minds, our little personalities, we shall become one with It. This is what is meant by "Thou art That". For we must know our true nature, the Absolute.

The finite, manifested man forgets his source and thinks himself to be entirely separate. We, as personalised, differentiated beings, forget our reality, and the teaching of monism is not that we shall give up these differentiations, but we must learn to understand what they are. We are in reality that Infinite Being, and our personalities represent so many channels through which this Infinite Reality is manifesting Itself; and the whole mass of changes which we call evolution is brought about by the soul trying to manifest more and more of its infinite energy. We cannot stop anywhere on this side of the Infinite; our power, and blessedness, and wisdom, cannot but grow into the Infinite. Infinite power and existence and blessedness are ours, and we have not to acquire them; they are our own, and we have only to manifest them.

This is the central idea of monism, and one that is so hard to understand. From my childhood everyone around me taught weakness; I have been told ever since I was born that I was a weak thing. It is very difficult for me now to realise my own strength, but by analysis and reasoning I gain knowledge of my own strength, I realise it. All the knowledge that we have in this world, where did it come from? It was within us. What knowledge is outside? None. Knowledge was not in matter; it was in man all the time. Nobody ever created knowledge; man brings it from within.

It is lying there. The whole of that big banyan tree which covers acres of ground, was in the little seed which was, perhaps, no bigger than one eighth of a mustard seed; all that mass of energy was there confined. The gigantic intellect, we know, lies coiled up in the protoplasmic cell, and why should not the infinite energy? We know that it is so. It may seem like a paradox, but is true. Each one of us has come out of one protoplasmic cell, and all the powers we possess were coiled up there.

You cannot say they came from food; for if you heap up food mountains high, what power comes out of it? The energy was there, potentially no doubt, but still there. So is infinite power in the soul of man, whether he knows it or not. Its manifestation is only a question of being conscious of it. Slowly this infinite giant is, as it were, waking up, becoming conscious of his power, and arousing himself; and with his growing consciousness, more and more of his bonds are breaking, chains are bursting asunder, and the day is sure to come when, with the full consciousness of his infinite power and wisdom, the giant will rise to his feet and stand erect. Let us all help to hasten that glorious consummation.

Vivekananda widely popularised Vedanta in the West, but his teachings represent only one particular school, called Advaita (literally "non-dualism," often called "monism". Vivekananda somewhat modernised the original Advaita doctrine of Shankara. The main antagonists of Advaita are the dualistic schools of Vaishnavism.

Vedanta is closely aligned with its sister school, Mimamsa (meaning "enquiry"). The two are often called Purva Mimamsa and Uttara Mimamsa, denoting the earlier and later schools of enquiry. The earlier Mimamsa deals with *dharma* and focuses on rituals, particularly for promotion to higher planets. Its main texts are from the *Karma-kanda* section of the Vedas.

The later school, Vedanta, deals with Brahman and derives much from the *Jnana-kanda* section, especially the Upanishads and Vedanta Sutras (also known as the Brahma Sutras). Members of these two doctrines are traditionally at loggerheads, but they are essentially complementary. Much of modern Hinduism, though philosophically grounded in its Vedanta, borrows its ritualistic practice from *Purva Mimamsa*.

There are ten principal schools of Vedanta. Of these, two are purely Advaita, one purely Dvaita and seven a synthesis of both. In Hinduism as represented in the UK, there are four important schools of Vedanta.

The Four Main Schools of Vedanta Represented in the UK

For any tradition to be accepted as a genuine Vedantic school, the founder must present commentaries on three scriptures–the Upanishads, the Vedanta Sutras, and the *Bhagavadgita*, collectively called the "*Prasthana trayi.*"

Below are the four main branches of Vedanta along with: its founding theologian, the *Sampradaya*, the main representative movement in the UK.

School	*Theologian*	*Sampradaya*	*UK Movement*
Advaita (monism)	Shankara	*Smartas*	Ramakrishna Mission
Vishishtadvaita (qualified monism)	Ramanuja	Shri Vaishnavas	Swami Narayana Mission
SuddhAdvaita (purified monism)	Vallabha	Pushti Marg	Shree Vallabha Nidhi
Achintya Bheda-abheda	Baladeva	Gaufdiya Vaishnava	ISKCON (Hare Krishna Movement)

Common Misunderstandings

Vedanta (and all Hinduism) is entirely monistic, believing only in the all-pervading world-soul, Brahman, rather than a personal God.

This *Advaita* philosophy is certainly popular, and offers a simple explanation of the many deities. Nonetheless, many theologians have considered God to be a person. He is not merely an anthropomorphic representation, nor are the various deities and *murtis* simply incarnations or representations of an impersonal Supreme.

Thus Hinduism includes both monism and monotheism. It is misleading to call the Abrahamanic religions, "the monotheistic traditions," implying that monotheism is absent from the Eastern traditions. Vedanta includes many monotheistic schools. They may accept the existence of many gods and goddesses, but strongly emphasise the pre-eminence of the Supreme Deity.

2

Purva Mimamsa, Advaita and other Schools of Vedanta

The Mimamsa (literally, enquiry) schools admit of *Shruti* (that which is heard, *i.e.* revelation) as a pramana. Usually, the word Mimamsa refers exclusively to the Purva Mimamsa school. The uttara Mimamsa school is more popularly known as Vedanta. A wide variety of texts, collectively called *Smriti* (that which is remembered), is taken as a lower authority that is valid when it does not conflict with *Shruti*. Other sources of knowledge accepted by both Purva Mimamsa and Vedanta are: *pratyaksha* (sensory perception), *upamana* (analogy), *anumana* (inference), *arthapatti* (postulation) and *anupalabdhi / abhava* (non-cognition/absence).

The first three are borrowed from the Nyaya-vaiseshika schools, but arthapatti and anupalabdhi are unique to the Bhatta school of Mimamsa thought. The Mimamsa school of prabhakara, called guru-mata, does not accept abhava as an independent pramana. Above all these, the veda stands as the supreme source of knowledge. The primacy concern of Mimamsa is textual exegesis, in addition to being a school of philosophy. Theology and religion are inseparable from philosophy in the Mimamsa schools. However, both Purva and uttara Mimamsa schools maintain that *Shruti* exists only to reveal that which cannot be known otherwise. Moreover, *Shruti* cannot deny a fact that is amenable to ordinary perception, *e.g.* no amount of repetition by the vedas that fire is cold will make fire cold. Thus, if there occurs a statement in *Shruti* that goes contrary to perception, it requires interpretation in a metaphorical or allegorical sense. Hence, there is a place for logical thinking based on perception and inference in these schools.

Purva Mimamsa interprets the vedas mainly as a set of injunctions (*vidhi*), with adjoining recital (*mantra*) and commentary (*arthavada*) portions. Thus, a statement, "he who is desirous of heaven should perform the jyotishtoma rite" is a *vidhi*, an injunction, and the rite itself is to be performed with the relevant mantras. The knowledge conveyed by this statement is not known by any other means, and its purpose is to impel the listener to action. According to Purva Mimamsa, this heaven is the highest salvation that is available to human beings, and thus the vedas are the sources of knowledge about both *dharma* and *moksha*.

The Upanishads and Brahmanas which relate to the said jyotishtoma rite are considered to be the arthavada, the explanatory commentary. The knowledge conveyed by the Upanishads is also not known by any other means, but the Upanishads are considered to be subordinate to the statements which impel man to action. Along with injunctions, there are also *pratishedhas*, statements which prohibit certain actions. The tradition of such textual exegesisis codified in the *Mimamsa sutras* of *jaimini*, with commentaries by *Sabara*, and sub-commentaries by *Kumarila Bhatta* and *prabhakara*. The Mimamsa that is taught in traditional vaidika pathasains in India today is based mainly upon the Sabara Bhashya and kumarila's Varttikas. This school is called Bhatta Mimamsa; the rival school called guru mata, following prabhakara's interpretation, is mostly extinct today.

Uttara Mimamsa, also called Vedanta, does not consider the Upanishads and Brahmanas as arthavada subservient to vidhis. Instead, they are seen to be sources of Brahman knowledge, addressed solely to those who seek *moksha*. The rituals enjoined in the vedas are applicable to the realm of *dharma*, but the one who seeks liberation does not merely desire a place in heaven; he is in search of ultimate Reality itself.

The Upanishads are viewed as those portions of the Shruti that address philosophical questions about Reality, here called

Brahman. This tradition of exegesis follows the *Brahmasutras* of Badarayana. Within Vedanta, there is considerable difference of opinion on whether the Upanishads enjoin anything at all. The non-Advaita schools consider some of the statements in the Upanishads to be injunctions. The most famous example is *Atma va are drashtavya: Srotavya: mantavya: nididhyasitavya:*. Most, if not all, non-Advaita schools would take this to be an injunction.

In contrast, the advaitin approach to this statement is to treat it as advice to the mumukshu, not as an injunction. This is because the Atma is an accomplished fact; it is not a result of any action and therefore meditation on the innermost Atma cannot be enjoined. In this respect, knowing the *Atman* contrasts with the heaven which is attained as a result of the performance of the jyotishtoma sacrifice. According to Advaita Vedanta, the veda addresses itself to two kinds of audiences-the ordinary ones who desire the transitory heaven and other pleasures obtained as a result of ritual sacrifices, and the more advanced seeker who seeks to know Brahman.

Thus, the Purva Mimamsa, with its emphasis on the Karma kanda of the vedas, is meant for the first audience, to help lead its followers along the way. However, the Vedanta, with its emphasis on the Jnana kanda, is meant for those who wish to go beyond such transient pleasures.

As a Mimamsa, Vedanta has a function of textual exegesis, with its uniquely Indian views on the origin, relevance and scope of revelation. In itself, any school of Vedanta can be considered to be a philosophy and also as a religion, there being no distinct line that can be drawn between the two, at least in the Indian context. Vedanta literally means "the end of the vedas." This can be interpreted in more ways than one.

From the textual point of view, the Upanishads, the source books of the Vedanta, are considered to be the end of the vedas. From a philosophical point of view, Vedanta is the fruit or the goal of all the vedas. The philosophical interpretation is more preferable in the tradition. This is because the vedic texts are considered to be eternal, revealed scripture, so that they have no definable chronological beginning or end. Advaita Vedanta is the oldest living school of Vedanta. It is also generally considered to be the premier school of Vedanta, and the word Vedanta is used synonymously with Advaita in most literature.

Vedanta bases itself mainly upon three sets of texts, called *Prasthana trayi*. These are the Upanishads (*Shruti Prasthana*), the Bhagavad Gita (*Smriti Prasthana*) and the Brahmasutras of Badarayana (*Nyaya Prasthana*). A working definition for a Hindu tradition to be called Vedanta is that it should have definitive commentarial texts on the three *Prasthanas*. Consequently, the following brief description overlooks important traditions like those of Kashmir Saivism and southern Shaiva siddhanta.

An essential identity between *Atman* and Brahman is upheld in *Advaita* Vedanta. The personality of Shankaracharya and the force of Advaita teaching is so strong that most post-Shankaran schools of Vedanta consciously define their doctrines against Advaita thought. A number of pre-Shankaran vedantins seem to have been proponents of *bhedabheda* (bheda+abheda, or *identity* in *difference*). The earliest post-Shankaran school of Vedanta is also one of ***bhedabheda***. This is seen from the commentaries of bhaskaracharya, which are still available, although the number of followers of this school is quite small. In the 14th century, Sripati pundita, a commentator from the virasaiva tradition, also identifies himself as a bhedabhedavadin. However, in general, the virasaiva school does not pay much attention to vedantic questions. Some early Advaita Vedanta influence on the monistic schools of Kashmir Saivism is also postulated. However, these schools base themselves upon an independent set of texts, namely the *Shaiva Agamas*, and do not consider themselves to be vedantic in origin.

A body-soul relationship between *Atman* and Brahman is upheld in the school known as *Visishtadvaita*. Here, the highest Brahman is considered as a "person" with a body consisting of souls and matter. Theistic devotion to this highest Brahman is held to be the sole means to final moksha. The Visishtadvaita schools split on the question of identifying the highest Brahman with Shiva or Vishnu, the Great Gods of Hinduism. The school of Srikantha regards Shiva to be the highest Brahman, and teaches a variety of Visishtadvaita. However, appayya Dikshita reinterpreted Sriantha's thought in terms of Advaita Vedanta, and consequently, this school has come to be known as *Sivadvaita*. Thus, only the Vedanta schools associated with Vaishnava religion have maintained an identity distinct from Advaita Vedanta.

Most people understand the word *Visishtadvaita* to refer to the Sri Vaishnava school of Ramanujacharya, which considers the highest Brahman to be Vishnu as Srimannarayana, and continues to have a distinct identity in southern India, with major centres at Srirangam and Kanchipuram. There is remarkable similarity between the teachings of Ramanuja and Sriantha, except that while the former insists upon the supremacy of narayana, the latter insists upon that of paramasiva. Many northern bhakti schools trace their thought to Ramanuja's tradition, through the person of ramananda, a Sri Vaishnava monk, who travelled extensively in the north and had many disciples. These Vaishnava monks also belong to the tradition of *tridandi samnyasa*, as compared to the *ekadandi samnyasa* tradition of the Dasanami monks. The *tridandi* ascetics carry a staff that consists of three sticks tied together, symbolizing the unity of three separate entities (God, individual soul and matter). The *ekadandi* monks carry a single stick, symbolizing the essential identity of Brahman and *Atman*.

Complete difference and dualism is taught in the *dvaita* school of Ananda*Tirtha* (also known as purnaprajna). This is a Vaishnava tradition, centered at Udupi in Karnataka. Ananda*Tirtha* was the disciple of an Advaita Dasanami monk named achyutapreksha *Tirtha*, but he completely rejected Advaita teaching. Because of this historical legacy, monks of the dvaita tradition continue to use Dasanami suffixes, especially *Tirtha*, and are *ekadandi samnyasins*, although they would not interpret their single staff as a symbol of Brahman-*Atman* identity. The gaufdiya vaishnavas claim to be affiliated to the dvaita tradition, but their teaching of *Achintya bhedabheda* is quite different philosophically. They have also affiliated themselves with the *tridandi samnyasa* tradition in recent times. And there is the devotional Vaishnava school of

Vallabhacharya, which is known as pushti-marga, and as *SuddhAdvaita*. Despite this name, this school should not be confused with Advaita Vedanta. The gaudiya vaishnavas have substantial following in Bengal, and the vallabha school in Gujarat. nimbarka, another Vaishnava teacher, taught *dvaitadvaita*, which is similar to *bhedabheda*. This is a Vaishnava school that has a small following in the plains of the Yamuna river.

Advaita Vedanta

The Advaita philosophy is not easy to explain briefly, and it is not the purpose here to repeat in a www few page what takes whole volumes for accomplished experts. But we will prvide you with a brief synopsis of the various aspects of Advaita Vedanta.

A very important assumption in all Vedanta is that man suffers from bondage in the course of his life in this world. This is said to be *samsara*, which involves being caught in an endless cycle of births and deaths. The quest therefore is to seek a way out of this bondage, to break the cycle of rebirths and attain moksha or liberation. The most important issues in Vedanta have to be understood with respect to what constitutes bondage and what constitutes liberation.

The Advaita school is of the view that *Jnana* (knowledge) of man's true nature is liberation. Bondage arises from ignorance (*avidya*) of man's true nature, and therefore removal of ignorance roots out this bondage. Liberation is therefore nothing more or nothing less than man knowing his true nature. This true nature is his innermost essence, the *Atman*, which is nothing other than Brahman. He who knows this, not merely as bookish knowledge, but through his own Experience, is liberated even when living. Such a man is a *Jivanmukta*, and he does not return to the cycle of rebirths.

Brahman: It may be noticed that at first glance, Advaita's solution to the problem of man's liberation does not seem to involve God as a Creator or a Saviour at all. If all that is required is to know one's own true nature, what role does God have to play in this universe? Advaita's answer to this issue is buried in the advaitic conception of Brahman. One is the view of the *Brahmasutra* that Brahman is at once both the instrumental and the material cause of the universe.

The *Brahmasutra* holds such a view because there is nothing that can be said to exist independent of Brahman. Is Brahman then just a name for a universal set-the superset of all things in this universe? Not so, because Brahman has been described as beyond all change, whereas the perceived universe is full of change. Still, this universe is said to have Brahman as the only cause. At the same time, to understand Brahman truly is to know. It to be devoid of parts and diversity, and beyond all causality/action. Such a conception of Brahman derives from the Upanishads, which say *sarvam khalvidam brahma*—all this is indeed nothing but Brahman—on the one hand, and *neha nanasti kinchana*-there is no diversity here-on the other. Thus, the conception of Brahman as a Creator in Advaita is a unique one, and directly relates to the Advaita views on causality.

Parinama and Vivarta: There are different theories of causality described by Advaita vedantins, but they are all agreed that Brahman is the sole cause of the universe, *i.e.* both the instrumental and the material cause of the universe. The axiom that the One Brahman is the cause of the many-fold universe is the foundation on which the entire system of Advaita Vedanta is based, and numerous efforts have been made over the centuries, to address logical problems arising out of it. This Brahman is also held to be eternal and changeless.

It is easy to understand Brahman as the instrumental cause of the universe. This view is not very different from the traditional perspective shared by almost all religions—some creator is usually credited with having created this universe. This creator is the instrumental cause of the universe. What differentiates the *standard Vedanta position* from such general theistic views is that Brahman is simultaneously also the material cause of the universe. In other words, creation is never *ex nihilo*, but proceeds out of Brahman. Itself, although Brahman remains unchanged.

Common-sense views of material causality always involve some kind of change. Thus, for example, milk is said to be the material cause of curds. However, in the process of curdling milk, the milk cannot be recovered. All we have at the end is the curds, the milk being irretrievably lost. This kind of causality involving change is called ***parinama***. There is another kind of material causality. For example, gold is the material cause of an ornament made out of gold.

In the process of making the ornament, the metal does not change into something else. It is only drawn into another form, from a lump to an ornament; the gold remains gold. This kind of causality is called *Vivarta*, where the material cause itself does not change into something else. The chandogya Upanishad makes very telling use of this kind of causality in its illustrations of how "Being" alone is the original cause (*sadeva saumya idam agra As it, ekameva advitiyam*), and how all perceived change is only in the realm of name and form, dependent on language (*vacarambhanam vikaro namadheyam*). The reality of gold is quite independent of what shape it is in.

Although Shankaracharya makes use of both kinds of causality (*parinama* and *Vivarta*) in his analogies, he denies that Brahman's role as the material cause of the universe involves any change in the essence that is Brahman. In the logical extreme, both *parinama* and *Vivarta* views of causality are deficient, as they presume a separate reality of the effect, apart from that of the cause.

Therefore, the most subtle arguments in Advaita Vedanta turn upon the *ajati vada* notion-that there is no real creation. *Vivarta* and *parinama* are both seen as convenient ways of describing causality, only if some provisional reality is conceded for the notion of creation.

Those who follow the *drshti-srshti vada* also maintain that Brahman is beyond all causality. However, most post-Shankaran authors, who teach in accordance with what is called the *srshti-drshti vada*, opt for a *Vivarta* notion of causality, as far as accounting for all creation is concerned. It should be remembered that the conception of Brahman as both the material and instrumental cause of the universe implies a very special kind of causality, one that is not similar to any other, and that cannot therefore be captured completely by any analogy.

It is as if Brahman has acted upon itself in order to produce this universe, that is full of change. Yet, the Upanishads abound with passages denying that any change is possible in Brahman, and indeed Shankaracharya denies that Brahman really acts. Brahman is also described as devoid of all attributes, along with passages that glorify Brahman as Ishvara, the Lord of this universe, with infinite attributes.

Nirguna and Saguna Brahman: To resolve such passages in the Upanishads, Advaita Vedanta maintains that really Brahman is devoid of all attributes, and is therefore known as *nirguua*. Brahman may be described as in the Upanishads, as Truth (*satyam*), Knowledge (*jnanam*), Infinite (*anantam*), or as Being (*sat*), Consciousness (*cit*), Bliss (*Ananda*), but none of these terms can be truly interpreted as attributes of Brahman as a Super-person/God. Rather, it is because Brahman exists, that this whole universe is possible.

It is because Brahman exists that man ascribes attributes to Brahman. However, Brahman's true nature cannot be captured in words, for all these attributes are ultimately just words. Hence, it is man's ignorance of its true nature that postulates attributes to Brahman, thereby describing It in *saguna* terms (with attributes). This saguna Brahman is Ishvara, the Lord, whose essential reality as Brahman is not dependent on anything else, and does not change because of the production of this universe. Therefore, Advaita holds that Brahman's own nature (*Svarupa-lakshaNa*) is devoid of any attributes (*nirguua*), while It is seen for the temporary purposes of explaining creation (*tatrastha-lakshana*) to be Ishvara, with attributes (*saguna*).

So much for *saguna* and *nirguua* Brahman. If Brahman cannot be held to have suffered any change because of creation of the universe, then what is the status of this universe? Since the cause does not undergo any change in the process of producing the effect, it is held that the cause alone is Real. The universe only partakes in reality inasmuch as it is to be considered as dependent on Brahman. Therefore the Upanishads say, "*sarvam. khalvidam. brahma.*" If the universe is considered to be independent of Brahman, then it has no real Reality, although the world of human perception can never reveal this truth.

This is simply because Brahman Itself is never an object of human perception. It is this characteristic of dualistic knowledge, derived from perception alone, that prompts the advaitin to call it *mithyajnana* (false knowledge).

Avidya and Maya: Why does human perception fail to see Brahman directly? Shankaracharya attributes it sometimes to *avidya* (ignorance) and sometimes to *maya* (the power to deceive). As the brhadaranyaka Upanishad puts it, "*vijnataram, are kena vijaniyat?*"-How is the Knower Itself to be known? It also stands to reason, therefore, that any effort at characterizing Brahman falls far short of Brahman. No words reach Brahman; how can mere verbal descriptions claim to describe It? Advaita now turns to the ancient technique of *adhyAropa-apavada* (sublation of superimposition) to explain this.

Thus, although Brahman is called the instrumental and material cause of the universe, Advaita tells us that this is only a preliminary view of Brahman, motivated by a need to explain creation of the universe. In order to understand Brahman, one has to go beyond this preliminary view, and understand Brahman in itself, not necessarily in relation to the universe. Then it is understood that the whole universe is only superimposed on the underlying Reality that is Brahman. To really know Brahman, one needs to sublate this superimposition, and look at the substratum (*Adhishthana*) that is Brahman. As for the exact nature of *avidya* and *maya*, later authors seem divided into two major schools of thought, namely the Bhamati and the Vivarana schools.

***Brahman** = Atman:* What then of the human self, the Jiva? It is here that Advaita comes up with the most radical answer, one that is unacceptable to all other schools of Vedanta. According to Advaita, what is called the universe is in reality not other than Brahman. Similarly, what is called the Jiva is in reality, the *Atman*, which is also nothing other than Brahman itself. The real Jiva is the *Atman*, which is unchanging, ever free, and identical with Brahman.

This is said on the basis of Upanishadic passages where the *Atman* is explicitly equated with Brahman. This equation of *Atman* with Brahman is also explained by means of *adhyaropa-apavada*. By sublating the superimposition of human shortcomings and attributes on the *Atman*, the pure *Atman*, the substratum, shines forth as Brahman Itself. The mani-fold universe and the individual self, which considers itself bound, are both superimposed upon that Transcendental Reality which is Brahman. Once the superimposition is understood for what it is, the individual is no more an individual, the universe is no more the universe-all is Brahman.

This doctrine of Advaita should not be misinterpreted to mean that the human self is in and of itself God, without any qualification whatsoever. Shankaracharya most emphatically asserts that such is not his intention. On the other hand, he is at great pains to point out that one who is desirous of moksha needs to overcome his human shortcomings in order to achieve full liberation. Shankara prescribes rigorous prerequisite qualities for the person who is to study Vedanta. These form the practical aspect of the effort to rise above and sublate the characteristics of the human Jiva, in order to understand the *Atman*/Brahman. The non-dual reality of the *Atman* is revealed to the intense seeker, as an experience that defies words. One might call it a mystic experience of Brahman, in which to know Brahman is to be Brahman. Thus, rather than being atheistic or non-theistic, Advaita Vedanta is meta-theistic: it points to the basic underlying Reality of all, including what humans call God, what humans call the universe, and what humans call human. This Reality is the unchangeable Brahman.

Tattvamasi: At this juncture, it is instructive to look at the advaitin interpretation of the chandogya statement *tattvamasi*, following Shankaracharya. This is one of the four statements that have become well-known as the Upanishadic mahavakyas, which equate *Atman* with Brahman. The four most important mahavakyas (one from each veda) are:

"*ayamatma brahma*" (mundaka)

"*tattvamasi*" (chandogya)

"*aham brahmasmi*" (brhadaranyaka)

"*prajnanam brahma*" (aitareya)

Shankara explains *tattvamasi* as follows. *Tat* is a common designation for Brahman in the Upanishads, while *tvam* (thou) addresses the student. The sentence states an equation of two seemingly different entities *tat*-that, and *tvam*-thou, by means of the verb *asi*-are. In general, Brahman (*tat*) is commonly understood as Ishvara (saguna Brahman), with an infinity of attributes, including the power of creation. *Tvam* is the individual who is bound, who is embodied, and who is in need of liberation. The difference between *tvam* and *tat* seems to be a matter of common knowledge for all individuals. What is the reason for the Upanishad to teach an identity then? An identity cannot be stipulated, even in infallible Shruti, if there is a real difference. Keeping in mind that Shruti is infallible, Advaita therefore concludes that really there is no ultimate difference between *tat* and *tvam*.

The identity expressed in a statement like *tattvamasi* is therefore held to be Real, and its realization constitutes the height of knowledge (*Jnana*). Direct experience of this *Jnana* is in fact moksha. It also follows that since this identity is not perceived normally, difference arises out of *avidya*, ignorance of the true nature of Reality. Since Shruti is superior to perception, this identity is indeed the supreme truth, all difference being in the realm of relative perception. If non-dualism is the true nature of Reality, why is this difference perceived in the first place? Given Advaita's basis on the non-dualistic scriptures, the perception of difference remains, in the final analysis, inexplicable.

This is labeled "*anirvacya / anirvacaniya* "in Advaita-something that can never be fully understood by the human mind. Since perception of duality presupposes *avidya*, no amount of logical analysis, itself based on this duality, will satisfactorily explain *avidya*. Hence, Shankaracharya is not much interested in explicating *avidya*, except to acknowledge its presence in all human activity, and in trying to overcome it to understand Brahman.

Vyavahara and Paramartha: This exegesis of scripture leads to the well-known advaitic doctrine of two levels of understanding: *Vyavaharika satya* (phenomenal or relative reality or just "reality", where duality is seen) and *paramarthika satya* (transcendental reality, or "Reality", non-duality). One important Upanishadic source for Advaita Vedanta's theory of two levels of truth is the analysis of the *Atman* as "*neti, neti*"- not this, not this. This is from the brhadaranyaka Upanishad.

This Upanishad also describes the highest state of the *Atman* in purely non-dualistic terms-"*yatra tvasya sarvam Atmaivabhut, tatra kena kam pasyet?..... vijnataram are kena vijaniyat?*"-Where the *Atman* alone has become all this, how is one to see another?..... How is the Knower to be Known? Most advaitins point to the quotation from the brhadaranyaka that immediately precedes this: "*yatra tu dvaitamiva bhavati,...*"- where there is duality, as it were,...-as the scriptural basis for saying that perception of duality is an appearance only, "as it were" and not the supreme Reality.

This rejection of all characterization as partial at best, and ultimately untrue, means that the *Atman* is beyond all duality, and all attempts to describe. It fail, because language itself presupposes duality. This *via negativa* approach is very much favoured in Advaita Vedanta. This emphasis on identifying the *Atman* with Brahman by means of sublating the commonly understood characteristics of each term, to affirm the real nature of the *Atman*, is central to Advaita vedanta.

Vishishtadvaita

Vishishtadvaita Vedanta is a sub-school of the Vedanta (literally, *end or the goal of the Vedas*, Sanskrit) school of Hindu philosophy, the other major sub-schools of Vedanta being *Advaita* and Dvaita. *Vishishtadvaita* (literally "Advaita with vishishtam") is a non-dualistic school of Vedanta philosophy. It is non-dualism of the qualified whole, in which Brahman alone exists, but is characterised by multiplicity. It is a school of Vedanta philosophy which believes in all diversity subsuming to an underlying unity. Ramanuja, the main proponent of Visishtadvaita philosophy contends that the Prasthana Traya, *i.e.* the Upanishads, the Bhagavad Gita, and The Brahma Sutras are to be interpreted in way that shows this unity in diversity, for any other way would violate their consistency.

Visishtadvaita philosophy of Ramanuja provided the philosophical basis for the establishment of Sri Vaishnavism and gave Vedantic backing to the brimming devotion of the Alwar saints and their composition of wonderful poetry and devotional songs in praise of Lord Vishnu. The succession of great Master-Expositors and spiritual giants of Vishishtadvaita school starts with the twelve Alwars, who left behind an imperishable legacy of Tamil devotional poetry in the form of 4000 songs, now called the Nalayira-prabandham.

The Visishtadvaita school of philosophy is probably the first school to interpret the Vedanta texts without being influenced by Buddhist thought.

Philosophers: The Visishtadvaitic thought is considered to have existed for a long time. The earliest known works of the philosophy are not available now. The names of the earliest of these philosophers is only known through Ramanuja's Vedanta Sangraha. In the line of philosopher's considered to have expounded the Visishtadvaitic system, the prominent one's are Bodhayana, Dramida, Tanka, Guhadeva, Kapardi and Bharuci. Besides these philosophers, Ramanuja's teacher Yamunacharya is credited with laying the foundation for what culminates as the Sri Bhashya.

Bodhayana: Bodhayana is considered to have written an extensive vritti (commentary) on the Purva and Uttara Mimamsas.

Tanka: Tanka is attributed with having written commentaries on Chandogya Upanishad and Brahma Sutras

Natha Muni: Natha-muni of the ninth century AD, the foremost Acharya of the Vaishnavas, collected the Tamil Prabandhams, classified them, made the redaction, set the hymns to music and spread them everywhere. He is said to have received the divine hymns straight from Nammalvar, the foremost of the twelve Alwars, by yogic insight in the temple at Alwar Thirunagari, which is located near Tirunelveli.

Yamunacharya: Yamunacharya, who renounced a kingship and spent his last days in the service of the Lord at Srirangam and in laying the fundamentals of the Vishishtadvaita philosophy by writing four basic works on the subject.

Ramanuja: Ramanuja is the main proponent of Visishtadvaita philosophy. The philosophy itself is considered to have existed long before Ramanuja's time. Ramanuja continues along the line of thought of his predecessors while expounding the knowledge expressed in the Upanishads, Brahma Sutras and Bhagavad Gita.

Vedanta Desika: Vedanta Desikan, one of the foremost learned scholars of medieval India, wrote more than a hundred works in Sanskrit and Tamil. All are characterised by their versatility, deep spiritual insight, ethical fervour and excellent expressions of devotional emotion in delightful style. His Paduka-sahasram is a classic example. He was a great teacher, expositor, debater, poet, philosopher, thinker and defender of the faith of Vaishnavism.

Vedanta Desikan is considered to be the founder of Vadakalai sect of Sri Vaishnavism.

Pillai Lokacharya: Pillai Lokacharya is considered to be the founder of Tenkalai sect of Sri Vaishnavism.

Visishtadvaita Philosophy: Every philosophy has to explain its position on Epistemology, Ontology, Metaphysics and Theism (*i.e.* the role and position of the Supreme).

Epistemology

Pramgas: Pram, in Sanskrit, refers to the correct knowledge, arrived at by thorough reasoning, of any object. Pramga (*sources of knowledge*, Sanskrit) forms one part of a tripumi (trio), namely,

1. Pramt, the *subject*; the *knower* of the knowledge
2. Pramga, the *cause* or the *means* of the knowledge
3. Prameya, the *object* of knowledge

In Visishtadvaita Vedanta, the following three pramgas are alone accepted as valid means of knowledge:

- Pratyakchha — the knowledge gained by means of *Perception*
- Anumna — the knowledge gained by means of *Inference*
- Chabda — the knowledge gained by means of *Shruti*

Chabda or Shruti Pramga occupies the highest position in matters which cannot be settled by Pratyakchha or Anumna.

Ontology: In more specific terms Vishishtadvaita conclusions may be briefly summarised as below. The Absolute Supreme Reality referred to as Brahman, is a Transcendent Personality with infinite superlative qualities. He is Narayana, also known as Lord Vishnu. He creates the other two members of the Trimurti, namely, Creator Brahma and Shiva, the Lord of Deluge.

Narayana is the Absolute God. The Soul and the Universe are only parts of this Absolute and hence, Vishishtadvaita is panentheistic. The relationship of God to the Soul and the Universe is like the relationship of the Soul of Man to the body of Man. Individual souls are only parts of Brahman. God, Soul and Universe together form an inseparable unity which is one and has no second. This is the non-duality part.

Matter and Souls inhere in that Ultimate Reality as attributes to a substance. This is the qualification part of the non-duality.

Metaphysics

Ethics: Souls and Matter are only the body of God. Creation is a real act of God. It is the expansion of intelligence. Matter is fundamentally real and undergoes real revelation. The Soul is a higher mode than Matter, because it is conscious. It is also eternally real and eternally distinct. Final release, that comes, by the Lord's Grace, after the death of the body is a Communion with God. This philosophy believes in liberation through one's Karmas (actions) in accordance with the Vedas, the Varna (caste or class) system and the four Ashramas (stages of life), along with intense devotion to Vishnu. Individual Souls retain their separate identities even after moksha. They live in Fellowship with God either serving Him or meditating on Him. The philosophy of this school is Srivaishnavism, a branch of Vaishnavism.

Comparison with Western Non-dualism: Baruch Spinoza, the 17th century rationalist philsopher, in his magnum work Ethics establishes the nature of GOD. Spinoza's pan-organistic God (*i.e.* God revealed as orderly nature) is comparable to Brahman (having the individual selves' and Universe as its body).

Spinoza makes the following propositions on the nature of God, in his work "Ethics". These positions closely reflect the Vishistadvaitic position on the nature of Brahman:

Proposition XI: God, or substance consisting of infinite attributes, of which each expresses eternal and infinite essentiality, necessarily exists.

Proposition XV: Whatsoever is, is in God, and without God nothing can be, or be conceived.

Proposition XVII: God acts solely by the laws of his own nature and is not constrained by anyone.

Proposition XVIII: God is the indwelling and not the transient cause of all things.

Proposition XIX: God and all the attributes of God are eternal.

Proposition XXX: Intellect, in function finite, or in function infinite, must comprehend the attributes of God and the modifications of God, and nothing else.

Dvaita

Dvaita (originally called Tattvavada), a school of Vedanta (the most widespread Hindu philosophy) founded by Shri Madhvacharya, stresses a strict distinction between God (Vishnu) and the individual living beings (Jivas). According to Madhva, souls are not 'created' by God but do, nonetheless depend on him for their existance. Satguru Sivaya SubramuniyaSwami explains dualism in his book, *Dancing* *with* Shiva, by stating that in dualism, God is seen as the efficient cause of the universe and not as the material cause. He is the potter causing the clay to emerge, rather than serving as the source of the clay itself.

Dvaita Philosophy: Like Ramanuja, Madhvacharya espoused a Vaishnava theology that understands Brahman to be endowed with attributes and a personal God, Vishnu. By Brahman, he referred to Vishnu, as per his statement "brahmashabdashcha Vishhnaveva" that Brahman can only refer to Vishnu. Madhva states that Vishnu is not just any other deity, but is rather the singular, all-important and supreme one. Vishnu is always the primary object of worship, and all others are regarded as subordinate to Him. The deities and other sentient beings are graded among themselves, with Vayu, the god of life, being the highest, and Vishnu is eternally above them.

Dvaita, or Dualistic philosophy, also known as Bheda-vada, Tattva-vada, and Bimba-pratibimba-vada, is the doctrine propounded by Ananda *Tirtha* (also known as Madhvacharya) that asserts that the difference between the individual soul or Jiva, and God, (Îshvara or Vishnu), is eternal and real. Actually, this is just one of the five differences that are so stated—all five differences that constitute the universe are eternal.

The Dvaita doctrine was summarized by Vyasa *Tirtha* as comprising nine tenets or Prameyas.

Impact of Dvaita Movement

- Madhva's Dualistic view, along with Shankara's Advaita or Non-dualism, and Ramanuja's Qualified Non-dualism, or Vishishtadvaita form some core Indian beliefs on the nature of reality.
- Madhva is considered to be one of the influential theologians in Hindu history. He revitalized an Hindu monotheism in light of attacks, theological and physical, by foreign invasion. Great leaders of the Vaishnava Bhakti movement, in Karnataka, for example, Purandara Dasa and Kanaka Dasa were strong proponents of the Dvaita tradition. Also, the famous

Hindu saint, Raghavendra Swami, was a leading figure in the Dvaita tradition.

- Madhva's theology heavily influenced those of later scholars such as Nimbarka, Vallabha, and Chaitanya Mahaprabhu. B.N.K. Sharma notes that Nimbarka's theology is a loose re-write of Madhva's in its most essential aspects. Vallabha even "borrowed without acknowledgement" a verse from Madhva's 'Sarva-shastrartha-sangraha'. The followers of Chaitanya claim a link to Madhva, though such a link is not historically tenable or theologically plausible.
- Madhva's singular contribution was to offer a new insight and analysis of the classical Vedantic texts—the Vedas, Upanishads, Brahma Sutra, Mahabharata, Pancharatra, and Puranas—and place uncompromising duality, which had been ravaged by attacks from Advaita, on a firm footing. Before Madhva, Nondualism was rejected by others such as the Mimamsa tradition of Vedic exegesis, and by the Nyaya tradition of classical logic. However, it was only he who could build a cogent alternative system of Vedanta that could take on Advaita in full measure.

Comparison to Mainstream Hinduism: Some teachings of Madhvacharya look different from mainstream Hinduism. One example is his doctrine of eternal damnation. Generally Hindus believe in the eventual salvation of every soul. Many of the doctrines in the Dvaita traditions resemble those of strict monotheism that is predominant amongst followers of Semitic religions. Dvaita provides a greater role to Bhakti than other schools of Vedanta. Followers of Dvaita believe in the supremacy of Vishnu over other deities including Shiva and do not believe in the Hindu concept of Trinity, Trimurti of Brahma-Vishnu-Shiva. Shiva is worshipped as a subordinate god (deva). Though this appears intolerant, it is because of the strong monotheistic belief in a non-Impersonal God unlike Advaita for which the identity of God does not matter as it is Nirguna. Historically, Dvaita scholars have been involved in vigorous debates against other schools of thought, especially Advaita.

Whereas Adviata preaches that *Atman* and Brahman are one and the same, which is not evident to the *Atman* till he comes out of a so-called illusion, Sri Madhvacharya puts forward the truth as Brahman (Vishnu) and *Atman* (soul) to be eternally different, with God always the most superior one. It is the same point that Madhvacharya reinforces in one of his doctrines... *"Yadi Namaparo Na bhaveth Sri Hari, khathamasya vasheth Jagathedhabhooth. Yadi Namanathasya Vashe Sakalam, Khatamevathu nithya suknham Na Bhaveth"*

The translation of the above is:

> *"If you feel there is no God, how do you explain as to why you cannot free yourself from the limitations on Earth? If you feel YOU are the one in control of everything (as Advaita preaches that Soul and God are one and the same), then how come you don't enjoy happiness always and are also subject to sorrow and pain (as God is supposed to be an eternity of happiness)?"*

3

Vedanta—Essentials of the Higher Values of Life

Some Basic Truths of Life: You have all come here for study. The traditional method of studying the Indian scriptures takes a minimum period of twelve years, and mind well, Saturdays and Sundays also working days and there were six hours of teaching and eight hours of self-study daily. In the first phase, they taught the Sanskrit language, grammar, *nyaya-shastra,* logic, etc. Then they used to teach the preliminary texts, the *prakarana granthas,* like Viveka-Chudamani, Tattva-Bodha, Atma-Bodha. Then only, the *prasthana-trayee,* the three main treatises, namely, Srimad-Bhagavad-Gita, the principal Upanishads and the Vedanta-Sutras which is also known as Brahma-Sutras were taught.

The normal period of study required in this traditional ancient *gurukula* method, was twelve years; but sometimes it was more, fifteen years or even twenty years, to study the intricate interpretation of words, their hidden meaning and significance. The different schools of thoughts may have different interpretations of the same combination of words. One such example is the combination *tattvam-asi.* One school may say it means 'thou art That.' Another school may say it is combined with the previous word and when separated it should read *atattvam-asi*, meaning exactly the opposite. Due to such intricacy of the Sanskrit language, a full-time study over a very long period is necessary.

Nevertheless, in this short period of teaching here, a brief outline of the Indian and the Western philosophies and the essence of the Indian scriptures is given to you here, and you will just get an entry into the Indian scriptures, so that you can pursue self-study after going back home. This serves the purpose of *sadhakas* and *jijnasus*.

The traditional twelve year study and passing the various examinations and getting the degrees of *shastri* and *acharya* is necessary for those who want to start their own *pathashala* or take up the career of a teacher or become a *purohit*. That is a totally different field of study in which is included a thorough knowledge of various sacraments, rites and rituals covering various ceremonies like naming the child, thread ceremony, marriage ceremony. This is known as *karmakanda*. A *purohit* is an important member of the Hindu society as no important ceremony can be taken up without him. In the past this was quite a lucrative profession and came in the family from generation to generation.

The Object of the Course: The study of the various paths of Yoga and of Upanishads and Srimad-Bhagavad-Gita provide practical guidance for Sadhana. In the West, there are theological seminaries for the Christians. But, for the philosophical study there is a separate department in many Universities where they teach the thoughts of the ancient philosophers like Plato and Aristotle as well as modern philosophers like Kant and Hegel. Like that, in every country there is some system for the study of both these aspects.

One may ask what is the purpose of all these philosophical texts? Why such a phenomenon called philosophy came into human social process of evolution? There are various branches of spiritual literature with relevance to day to day life. But what is the relevance of the study of philosophy? The purpose of my elaborating these details in connection with philosophical study is to ask you, "What is the object of this study of Yoga and Vedanta? What do all the various Eastern and Western philosophers seek? What do you expect to obtain by this study? What is the purpose of the spiritual knowledge?"

All the secular sciences left the phenomenon of man unexplained, left the origin of man unexplained. So the thinkers who wanted to know about this, probed into the mystery of man and his origin and thus came into existence the subject of philosophy.

What is the subject, scope and area of study of philosophy? It is twofold:

(1) It is a study of man. What is the phenomenon of life? What exactly is he constituted of? What is his origin? What is his ultimate destination?

(2) It is a study of the universe in which he lives. What is the nature of the universe?

Philosophy seeks to understand the phenomenon of life. So it has to have a minimum of two dimensions: the beings who live the life and the universe in which they live. Thus, man and the world become the two minimum reducible unavoidable subjects. A deeper study of these two factors leads to the study of a third factor: the origin of the man and the world. What is the origin of the universe? From where this world, sun, moon, planets, stars, galaxies, nebulae and all that have come? Who governs the phenomena of various natural cycles like day and night, different seasons and the phenomenon of rains which is just sufficient to support and sustain the life? What is the origin of man? There must be some source. So a third factor enters in the philosophical analysis and study. This third factor, the unknown source or origin of man and the universe, becomes an important part of the philosophical thought.

One may justifiably ask, "What do I gain by studying this unknown factor?" It must have some practical utility, some valid value. Otherwise, it would have been a for the time-being subject and disappeared soon. But it has persisted for a very long time and is growing and expanding, it is being enhanced and augmented generations after generations. This means that it must have some intrinsic and intriguing value in it.

Suppose you start with a pragmatic and logical approach and come to a certain conclusion that this must be the origin and source of the universe. In fact, most of the conclusions of the Western philosophical thought are derived in this way. They show tremendous speculative heights and a very very amazing great depth and details, incisive analysis, reasoning, logic and wisdom. They theorised their conclusions. They did not accept the previous conclusions and often disproved them by logical arguments and reasoning. They have a quest of perfection and always try to improve upon the existing theories. In this way, great philosophical systems have been evolved in the Western world by great scholars, thinkers and philosophers.

The Role of Saints: The Eastern philosophy is developed in a totally different manner. Our entire philosophical system is based upon and has been derived from the already existing knowledge which is found to be existing from ancient times, beyond the known history of humanity. God only knows when it originated. Its time is not traceable.

All our philosophical systems that are existing now have originated from the in-depth study of that Knowledge which was already existing. Our ancient seers and sages had already presented us certain conclusions about the basic philosophical questions and processes and sealed it ultimately with the seal of indisputable authenticity and genuineness of *Aparokshanubhuti*, their direct personal Experience of the Truth. In other words, the proof and validation of their conclusions was not the logical reasoning, analysis and arguments but this personal Experience. That is the uniqueness of our philosophical system.

In all the countries, in all the races, in all the cultures of the world we observe a unique phenomenon. Quite apart from the philosophers and scholars, there is a separate, unique and distinct category of people, the saints. Every race has produced some saints. Many of these saints were not very highly educated; some of them had even never gone to school, they were illiterate. Yet the world remembers them. Even in the modern world of science and high technology many people study their lives and teachings, and draw inspiration from them though they were neither philosophers nor worldly people. Even after thousands of years their lives and teachings have remained a source of inspiration.

In the beginning came the worship and sacraments, the *karmakanda*. They realised, "This is not enough. Even if we worship daily, even if we perform all the sacrifices and oblations, all the *yajnas* and *yagas* and all the *samskaras,* we can never be free from sorrow and grief, pain and sufferings, old age and ailment, tension and anxiety, fear and fright, fights and quarrels, diseases and death. We have no lasting peace of mind. So we must find out something higher, something which is beyond this and which will really free us. We must find out the state which is *dvandvatita* and *trigunatita,* transcending the pairs of opposites and the three *gunas.* We must find the state beyond sorrow, the state of eternal peace. Let us contemplate on That only."

Ultimately, in some wonderful moment in the history of India, in the history of mankind, some of these intrepid and gallant seekers, explorers of the unknown and unseen realm, came to experience the Ultimate Truth, the Supreme Reality. They had the direct Experience. They shouted in wonder and amazement, shouted in joy, "Ah! This is That! This is the Light of lights! This is the Reality, the Truth!"

Aparokshanubhuti: What does the direct Experience mean? All the knowledge a human individual gathers in this material world is through the perception of senses. If you have a good hearing faculty you gather knowledge through all kinds of sounds; you can distinguish all types of sound, all types of speech, words, etc. Similarly, if you have a perfect faculty of seeing, your cognition of the entire world is through sight. In the same way, you are able to perceive heat and cold, soft and hard through the sense of touch. Similarly you perceive the world through the sense of smell and the sense of taste. So, whatsoever you know about the world right from the birth onwards, is through the five senses of seeing, hearing, touching, smell and taste in the form of *roopa, shabda, sparsha, gandha and rasa.* (Appearance, speech, touch, smell and taste) These are the five dimensions of your cognition.

These are five types of experiences through which you perceive the whole world and acquire all the knowledge. So this is indirect knowledge, it comes through one of the senses or a combination of them. If any one of these senses is absent, that part of perception is absent. A person born blind has no perception of colour, form and appearance. That aspect of the world does not exist for him at all.

In direct Knowledge the senses play no roll. *Aparokshanubhuti, the* direct Knowledge arises only when the senses are completely still. The senses are useless for direct Knowledge. As a matter of fact the senses are the main obstacle in direct Knowledge, they come in the way of the direct Knowledge. It arises without the medium of these senses, when the senses are totally still, the mind is desireless and when the seeker enters a very subtle state of higher consciousness. One just experiences in the depth of one's own consciousness, one's own real Self, transcending the material world and its cognition. That is the direct Experience. That is the *Aparokshanubhuti.*

In some wondrous era in our ancient history, the seers had the great Experience and came face to face with the Truth, came face to face with the Ultimate Reality. The moment they entered into the direct Experience, they were amazed, they were struck-dumb; because, there was a total transformation of their whole consciousness and they became filled with the Light. The darkness of ignorance vanished instantaneously, all doubts, all questions vanished forthwith. They experienced the total Knowledge; everything became clear straightway; no more any questions, any doubts, remained. That Knowledge is known as *Brahma-jnana* and the knower is known as a *jnani*. Thus, whatever is to be known became known to them. The origin and the source of the universe and the man, also became known to them. They became the very embodiment, the very personification of Knowledge itself.

Vedanta: The records of these purely transcendental experiences comprise and constitute the contents of the end portion of *Vedas* known as *jnanakanda* or the *vedanta*. Some of these seers have very compassionately narrated how they attained It, what were all the Sadhana and *tapas* they did. So Vedanta is the most important and valuable portion of the Vedic wisdom heritage. It forms the very basis and foundation of spiritual life. It is the source and origin of all the Indian philosophical systems, whether it is the Absolute Monism of Sankaracharya, or Qualified Monism of Ramanujacharya or

Pure Dualism of Madhvacharya or Dual-nondual Doctrine of Nimbarkacharya or Pure Dualism of Vallabhacharya or *Achintya Bhedabheda* of Gauranga Mahaprabhu. All the great *acharyas* based their philosophical systems on different interpretations of *vedanta*. All these philosophical systems, the schools of thoughts, are known as *vedanta*.

There are many schools of philosophy in the East, Far-East and the West. But the main fundamental subject, namely, the purpose of human life is more explicitly and emphatically spelt out in the East. Before making a detailed study of the philosophical systems, let us inquire into some basic questions: What is our purpose of this study? What do we obtain or gain from such study? What are their fundamental findings and teachings? What do these seers want to convey to us?

When we try to probe this particular question, we find that what they want to convey to us ultimately, is something so very apparent, so very absolutely clear and plain that nobody will ever question it a second time. They have given us some plain facts, some home-truths of life. The way by which they reached them at that time must have required a great deal of thinking, a great deal of contemplation, a great deal of analysis and speculation and all that. We should be very grateful to them, for they have proclaimed them and very clearly put before us the conclusions they arrived at. They have spared us all the difficult spadework, all the agony. Now it is like reaching Mt. Everest by an aeroplane or helicopter instead of reaching there by one's own effort, by one's own self, slowly trekking the long route, climbing the extremely difficult mountains and making the final attempt to reach the summit again and again. Whatever is necessary for acquiring the Knowledge, our ancient seers have already done somehow and have clearly laid down the path for us. They have clearly derived the ultimate Truth. We have simply to follow their path.

The Basic Truths: Now I will tell you something that I usually tell on the last day as a concluding message. Without telling that I am not closing my message. But something prompts me from within to tell this today itself. Once you know it, your entire approach to your study, to your life, will be changed and it will take a new *bhava,* a new viewpoint altogether. The ultimate highest discovery of the ancient sages and seers is recorded in the *Upanishads,* which reveal the great truths and teachings. It is the great grace, compassion and blessedness of our ancients that some sages might have thought: 'In the times to come, who is going to study this very abstract Vedantic philosophy, understand it and acquire thorough knowledge, ponder and contemplate on it and attain Illumination?' They must have envisaged that the whole world will be career oriented and money will rule the entire life. All the time and energy of people will be spent in earning money and keeping the pot boiling at home.

Though the teachings of Upanishads is the highest and the greatest invaluable treasure of the mankind, very few people will be able to take interest in it. So the quintessence of the entire wisdom teachings, knowledge and spiritual experiences of the Upanishads is given to us, to the entire mankind, in a very concise yet very simple, lucid and interesting form in the Srimad-Bhagavad-Gita. In this amazing and epigrammatic poem, all the truths are included. If we try to keep in view what basic, fundamental yet practical and pragmatic home-truths Bhagavad-Gita gives us, we can make our life based on them and enrich it.

Just imagine in eighteen chapters, in seven hundred verses only, the quintessence of the entire teachings, the wisdom knowledge of all the Upanishads, the highest spiritual experiences of so many illumined sages and seers, are treasured in such a simple and easy-to-understand way! Some of these verses are the preamble, the prologue and the narrative and the actual number of verses giving the intricate teachings is much less. It is amazing how in such a concise treatise, in such a brief discourse everything necessary for our uplift and supreme welfare is given to us!

If we try to keep in mind the basic and fundamental yet practical and pragmatic home truths this wisdom teachings of Bhagavad-Gita gives us, and if we base our life on them, we can attain our highest good. We cannot afford to ignore these basic truths. Unless our life is based on these truths, our life will be wasted. This is one interpretation of the implication of these truths.

Truth is God: Another implication of the fundamental Truth is: It is the ultimate Reality. That alone is the real Truth and compared to That, everything else is untrue. All other things fail to fulfil the test of genuineness and authenticity of Truth which That fulfils completely. So the definition of that ultimate, eternal Truth is *Tat Sat.* This is another implication, the highest one, of Truth.

There are some lesser implications also. The facts of life which no one contradicts are also known as truths. There are two views about this: there are certain facts which are so self-evident that there can be no dispute about them. These are known as *svatah-siddhah* truths. There are many such truths elaborated in the Upanishads, and the Gita has reproduced them in a simple, lucid form.

Such truths are narrated in more interesting, very simple, very easy-to-understand story form in the Puranas. In the Puranas a truth is woven in a very absorbing big story and many people may miss it. The narration is so interesting and elaborate that many people take note of the stories and bypass these truths and the preaching.

Certainty of Death: Whereas Bhagavad-Gita puts them in such a straight, direct, undiluted and concise form that no one can ever miss them. One such truth that all of you must always remember and which you cannot afford to forget even a single day of your life is proclaimed at the very beginning of Lord Krishna's teachings in Bhagavad-Gita. What is that? That great truth is: *The absolute certainty of death.* In one categorical statement He says, *Jaatasya hi dhruvo mrityuh* (Certain is the death for the born. II-27). For everyone who is born, death is certain. How many people have really pondered over the importance of this truth? You should ponder: 'Why this has been so categorically proclaimed?

In what way this concerns me? How should I relate myself to this great truth? Why *Jagadguru Purnavatara* Bhagavan Krishna told this to Arjuna?' He told this to Arjuna in order to bring him out of delusion, in order to take him out of nervous break-down, in order to bring him out of the weeping condition. He asks, 'For whom are you weeping? If you don't kill them, do you think that they will live for ever? Even if you don't kill them, one day they will die; die they must, in one way or the other. Even at this moment many are being born and many are dying. This is the plain truth of life. Why don't you realise it fully? Why don't you take it calmly? What is the use of breaking your head for something which is inevitable, for the truth which cannot be changed or altered?'

What was Arjuna's delusion? He thinks: 'All these people will die in the battle-field.' All of them had to die, death is inevitable for all of them. But Arjuna thinks: 'If he does not fight they will not die, they will live.' Krishna says: 'What a foolish notion! Whether you fight or not, all are going to die and you too will die.'

What is the lesson? The lesson is: If we do not recognise and realise the certainity of death very plainly and clearly, and if we do not accept this truth in living our daily life, we are also living in delusion. We do not want to face the hard reality of life. If we are living our life setting aside this truth, we live life in a state of ignorance and delusion. The root cause of all sorrow is this delusion. When someone near to us dies, we are dismayed and despaired as if something most unexpected has happened. But it is not at all unexpected. Lord Krishna clearly says: 'Everyone who is born has to die.' The time of death is most uncertain; but sooner or later everyone has to die. This truth is based on the actuality and we must accept it and always remember it.

Let us now see the constructive dimension of this truth. In what way it can be useful to us? If we recognise this first truth so forcefully brought out by Lord Krishna in Bhagavad-Gita wisdom teachings and if we always keep it in mind, we will never postpone anything that is important, we will never postpone our duties, we will never say: 'I will do my duty later'. Who is sure of tomorrow? Then how can you say: 'I will do it tomorrow?' It is said, *alone you come and alone you go back*. It is so correct from one point of view. But if we look from another point of view, one is never alone. From the very moment an individual *jivatma* is born, it is never alone; it is always accompanied by the Lord of Death. The whole life is a slow process of death, a constant march towards death. With every breath, with the passing of every moment, we are moving closer to death. You are never alone; the shadow of death always accompany you. The Lord of Death takes away your life every moment, little by little. At the last moment, he takes away the *pranas* and goes away and the body becomes lifeless.

If you recognise this truth: One day this life will terminate, one day I will die, I will not be here. I must ponder over this and resolve: 'Whatever I have to do, I must apply myself to it now itself, right at this moment; I will never postpone it to tomorrow.' Now you cannot be a loser. You are free from the delusion of tomorrow. Guru Maharaj used to say: Never say I will do it tomorrow. Tomorrow is the most deceptive word. There is no such thing like tomorrow. If you postpone something on tomorrow and go to sleep today; but when you wake up, it is already today and not tomorrow. God's ration is *today* only; tomorrow is our concept.

Past is finished, it has slipped from your hands and you cannot recover it again. There is no use projecting your important work and duty in future, because it has not yet come, it is not in your hands. What you have in your hands is *now,* the present moment only. Remember this always, every moment of your life.

You should always remember the death; remember that this life must terminate one day and you have to leave this world. Therefore resolve with firm determination: 'I will attain whatever I have to attain, whatever the goal of life I have. Right now, from this moment itself I shall engage myself in it, *I shall strive for the goal of life, right now.* I will attend to all my aspiration, duties and obligations, my *kartavya karmas.* Let me apply myself to it right now calmly and quietly, without tension or anxiety. I shall not postpone it, I shall not show any laziness or lethargy.'

God has wonderful sense of justice and has given to all of us enough time. Everyone has twenty-four hours a day, the same ration to everyone without any discrimination. Whether you are a professor or doctor, a big man or a beggar, you have the same ration. If you plan your time properly, there is time for everything, everything can be achieved in this life itself.

There is an English writer called Arnold Benett. He has written a wonderful book, 'How to live on 24 hours a day?' It implies that no one really lives for all the 24 hours a day. We fritter away little, little moments from morning to evening. This book proves that if you can make use of those moments which are unnecessarily wasted on trifle, irrelevant things, you have enough time for everything.

So, the very first truth proclaimed by Bhagavad-Gita is of great importance. You must live with full awareness that one day the life must end in death. Therefore, whatever you have to do, should be done right now. Start doing it right now. You cannot afford to remain idle and waste away the precious time. Time wasted is the time lost for ever, you can never get it back, you can never use it. Time does not give you a second chance. Whatever you have to do, whatever effort and exertion you have to put forth to achieve the goal, you have to do it right now. In this way, you will be able to achieve what has to be achieved.

The World is an Abode of Unhappiness: Now arises the next question: What it is that diverts us away from our goal? What is it that deceives and dupes us and makes a fool of us? What is it that makes us repent and regret that I have not done what I should have done? When the time of departure comes, you are very sad thinking: 'I have not been able to do what I ought to have done, I have not been able to become what I wish I had become.'

It is the deceptive attractions of the sense objects all around us. They entice us saying: 'Come, come! You will get all the happiness from us'. Some beautiful sight says, 'You will get happiness from me.' Some pleasant, attractive sound, smell, taste or touch says, 'Come, come! You will get happiness from me.' So, all our time and all our energy are consumed by numerous attractive objects of this wondrous supermarket of

Maya-bazaar. Your attention is scattered through the five holes of senses.

You do not realise that this is only a delusion and there is not an iota of real happiness in these objects. It is only some momentary pleasant nervous sensation. It is not a real happiness. Such pleasant experience is purely a gross, primitive, biological process and a phenomenon common to all the animals. When a sense object comes in contact with the related sense organ, the tactile nerve catches the sensation and sends a message to the central nervous system which passes the message to the brain and the brain feels a momentary pleasant sensation. You are now hooked by this sensation, you are enslaved. All these pleasant sensations are momentary and depend on the proper functioning of the apparatuses. Even if any one of them, a sense organ, the connecting nerve or the receiving centre in the brain is impaired, there is no sensation pertaining to that sense faculty. If you have lost the sensation of taste, you will not be able to experience any taste, even if the most delicious dish is given to you. If your optic nerve has become week, the most beautiful sight will have no impact on you even though your eyes may remain open.

How long is this experience of pleasant sensation lasts? You are eating your favourite tasty dish. So long as the food is in the mouth there is the pleasant sensation. But once it goes in, it is all over. In a flash of a moment the sensation disappears. But who thinks about this? If you ponder over it, if you analyse it impartially, you will realise that it is only a purely biological nervous sensation and this cannot be called real happiness. But we are duped by it and we interpret it as happiness and we waste our life in chasing the sense objects. Can there be a greater folly? It is a great blunder. It is not wisdom. There is no happiness here. On the contrary He who created all these, says:

Ye hi samsparshajaa bhogaa duhkhayonaya eva te;

Aadyantavantah Kaunteya an teshu ramate budhah.

[The enjoyments that are born of contacts are only generators of pain, for they have a beginning and an end, O Arjuna! The wise do not rejoice in them. B.G. V-22]

He makes it quite plain that all the enjoyments arising out of the contact of any sense object are the source of sorrow and pain only, there is no happiness in them. This is the second great truth given by Lord Krishna in Gita. This is a scientific and analytical way of looking at the problem.

He puts this truth in a general form also. He says, 'O Arjuna! Do you know what is the nature of this world? It is *duhkhaalayam-ashaashwatam* [the place of pain and is non-eternal. B.G. VIII-15]. This world is a world of transitory objects with names and forms. The transitory objects cannot give you lasting, eternal, real happiness. This world is an abode of misery and sorrow.'

In a negative way, He describes the characteristics of this world as: *Anityam-asukham lokam imam* [this impermanent and unhappy world. B.G. IX-33]. This world is characterised by a total lack of happiness. The senses are *duhkhayonaya eva te*, generators of pain. Again and again He reiterates this great truth that there is only pain and sorrow that you can obtain from this world.

Buddha's Four Truths: When Buddha attained the Illumination, he discovered four great truths pertaining to the human life:

- Life is full of suffering;
- The cause of the suffering is craving, desire (*trishna*);
- The possibility of complete cessation of the suffering by destroying desire; and
- The way to bring about the ultimate cessation of the suffering.

The first step is to become convinced that this world is an abode of sorrow and suffering. The next step is to know the cause of the suffering. Once we know the cause of the suffering, there is an unending hope for the mankind. Because, once you remove the cause the suffering will end. The suffering is an effect. Once the cause is removed the effect does not arise at all, it is automatically eliminated. Once you know the cause, you can find out by one way or the other, some device to remove it. The first discovery everyone knew; but so long as they do not know the cause of the suffering, people live with it thinking I can be happy with the coexistence of sorrow.

The third discovery is also very important. All these sufferings are not unavoidable or inevitable. They are not permanent. If the sufferings were eternal, if there had been no possibility of ending the sufferings, the question of finding a way to end them would have not arisen at all. But Buddha discovered that we can put an end to all the sufferings. He assured us that this is certainly possible.

Lastly, Buddha discovered the way to end all the suffering and putting completely an end to all the suffering. It is called The Noble Eight-fold Path. Like the eight-stage path to Liberation given by Maharshi Patanjali in his Yogasutras, Buddha has also discovered an eight-fold path to cessation of suffering: right views, right intention, right speech, right action, right livelihood, right effort or endeavour, right mindfulness and right meditation. Like in Yogasutras, Buddha also gives minute details of each of these. The first one is Right View and Buddha gives ten aspects of it.

Today I am telling this to reiterate that death is inevitable. Therefore, be alert! Be up and doing. Apply yourself quickly to whatever you have to do. Fulfil all your obligations and duties. *Be up and doing on the path of Liberation, on the path of Divine Perfection, the great grand, glorious Goal of human life.* You have to apply yourself to that Goal right now. What is that comes in your way as an obstacle? The delusion of happiness here. Convince yourself firmly that there is no happiness here and there can never be happiness here. This world is petty abode of *janma-mrityu-jaraa-vyaadhi-duhkha* (the evils of birth, death, old age, sickness and pain. Gita XIII-8), an abode of sorrow, sufferings and unhappiness only.

You are enslaved by the sense objects only if you think that there is some pleasure, some happiness in them. But once you discover their true nature, you are free. But once you are convinced that there is no happiness here, you are no longer deluded, you are no longer made fool of by the sense objects, you are wise. Through all these attractions you can remain untouched and unaffected by them. You will say: 'Yes, I know your nature. You can no longer fool me.' You become established in *anasakti,* total detachment. Even if you are surrounded by hundreds of attractive objects of pleasures, you are not perturbed. You know their real nature and so cannot be deluded. You go through your life as a master of the situation and not as a slave to every attractive sense object of name and form. You are the in-charge of your life.

If you have realised these truths, you become totally detached. Now you are steadily, serenely advancing to the Goal, absolutely unaffected by the sense objects. The basis of your detachment is discrimination and wisdom. You know that there is nothing in them; now they cannot enslave you.

A mother was frightening a little boy by covering the bush with a black cloth and was saying: 'Don't go there. A devil is hiding there.' The father thought that she is unnecessarily frightening the child, this is a bad psychology. The son will afterwards be always afraid. So one day he took the child near the bush and removed the cloth and said: 'Look, there is no devil here.' The child told his mother: 'Now you can never make me afraid of that black cloth. I know there is nothing in it. I know its real nature.'

Like that once you know the real nature of the world through constant discrimination, it will never be able to dupe you, delude you. You will no more be afraid of it. Now you can proceed straight towards the target like an arrow shot from the bow.

This is the second truth about the life about which I have to tell you. All the philosophers tell us that our tenure here is temporary, we are not residents of this earth plane, we are just the travellers. Our life is a journey with a definite destination. The important point is that you should always remember the Destination, the Goal. So many people go from here to the Badrinath shrine day after day. The journey is very difficult, arduous and full of discomforts. But they have only one idea fixed in the mind: 'I will reach my destination sooner or later. I will have the great fortune of having *darshan* of Lord Badrinath.' There is no thought about the discomforts and hardships.

The Divine Destination: The next great truth proclaimed by the Upanishads and Bhagavad-Gita is about our Divine Destination. Your destiny is to transcend the sorrow and suffering, pain and misery, defects and imperfections and to attain the state of glorious Perfection that is characterised by Bliss-absolute, Peace-profound. Perfection, Peace and Bliss. This is your birth-right. You are given this human birth only to attain that state. You are the heir of that profound Experience of unalloyed, unmixed, hundred per cent pure Bliss, a distilled quintessence of absolute Bliss. That Bliss, that Peace, that Happiness is imponderable.

It does not come from any source, It does not depend on any object. That Bliss, that Peace, that great sense of Perfection, that Light is known as Brahma, God, Bhagavan, Paramatman, Allah, Khuda. We call It the Kingdom of Heaven or the Kingdom of God. You may call It by any name, no matter what name, label or appellation you give It, It is that great Experience absolute where all the sorrows disappear, where all the doubts vanish, where all the questions disappear. It is transcending them all, It is transcending all sorrow, pain, grief, suffering, delusion and bondage. You are in a state of absolute freedom and fearlessness.

The Goal of human life is That. You have full potential for attaining It. It is your birth-right. You are given this human birth to attain That. That Experience is called *Kaivalya Moksha Saamraajyam*. That is to be kept as your Goal. Your efforts to attain That should never be postponed for *Jaatasya hi dhruvo mrityuh* (Certain is the death for the born. B.G. II-27). You should not be tempted and diverted into side-channels. There is nothing anywhere else. They are like water of mirage. You should know and remember always, *duhkhaalayam-ashaashwatam* [the place of pain and is non-eternal. B.G. VIII-15]. There is not an iota of real happiness here.

All these are fundamental truths. All the different schools of Vedanta whether it is the Absolute Monism of Sankaracharya, or Qualified Monism of Ramanujacharya or Pure Dualism of Madhvacharya or Dual-nondual Doctrine of Nimbarkacharya or Pure Dualism of Vallabhacharya or *Achintya Bhedabheda* of Gauranga Mahaprabhu, all of them declare these truths. They all tell this is a passing world; one day you have to die. You do not belong to this world, you are only a traveller. The human life is a great journey towards that grand Goal. It is not to be wasted after petty things. If your life is properly directed and properly used, *you can attain That in this very life*. You can attain Immortality, eternal Bliss and Peace at the end of your life. This is the quintessence of the Indian Philosophy. Contemplate on this. Don't be lost in nomenclature and unnecessary details. *You should know the heart of philosophy: You are Divine, you are the Immortal Atman. You are given this human birth to regain that Experience of your lost Divinity. Do it now. Do not postpone.*

The message of the day is:

- Remember always the death.
- Do not fall prey to the attractions of petty things.
- You are given this human birth to regain your lost Divinity. The goal of life is to attain It.
- Do it now. Be up and doing. *Uttishthatha Jagrata.* Awake, arise! Do not stop till the goal is reached. Keep on, keep on, keep on.

God bless you all.

The Golden Thread of Vedanta

The Divine Life Society was founded by worshipful Guru Maharaj Swami Sivanandaji, in order to proclaim to the whole world of the modern times, the way of living one's life so as to be able to attain, experience and become established in the Divine Perfection which is inherent in each one of the human being, as his unalienable, true, eternal nature. All of you have an apparent human personality which is only the external facade of your Reality.

This human personality is not the Reality according to the Vedanta. You have to know this because you are participants in the Yoga Vedanta Academy Course. So you must know what Vedanta has to say about your life, about this world and about everything pertaining to life in this world.

You have come here as *jijnasus*. *Jijnasus* are those who are keenly eager for attaining *jnana,* the Knowledge. *Jijnasus* attending this course have a special uniqueness, a special distinction. You are *jijnasus* with a difference.

Apara-vidya: What is that difference? The difference is: All other seekers of knowledge are trying to acquire *apara-vidya,* the worldly knowledge; whereas you seek to acquire *Para-vidya*. This is a basic, fundamental difference. *Apara-vidya* enables you to get, keep, possess and enjoy objects of this world. But Vedanta says, this world is only a temporary dream. *Apara-vidya* enables you to get distinction, name and fame, and the objects you can possess, keep and enjoy.

The World is Transitory: But all these only for a brief while; because the whole world is transitory, subject to decay, subject to ultimate dissolution. Everything here is unstable, nothing here endures, nothing here is lasting, nothing here is permanent. Therefore this world is not having any permanent validity. What is the meaning of everything? *Aabrahmaa sthoola paryantam*. From a blade of a grass up to the position of *Brahmaa*, the creator; *Brahma-loka*, the greatest, the biggest status in the cosmic scheme of things—everything is worthless, valueless. This is what has been declared by the great sages who have known the impermanence of all the things. So I have said that your outer personality is not the reality according to the *Vedanta*.

The Reality: Why did I say according to the Vedanta? Vedanta gives a specific definition for Reality. What is truly, authentically and genuinely Real? You have to try to understand with a very subtle intellect, what Vedanta wants to convey to you. That alone can be called Sat, really true, the Reality which is beginningless and endless, which is ever present, ever *is*, in all the three periods of time. It always existed in the past, It does exist in the present and will always exist in the future as well. It never will cease to be. There is a further qualification: That is the Reality which abides always, in all the three periods of time without undergoing any change, It always remains the same. Nothing can alter It, nothing can change It. That is something which is always in Its Fullness, in Its Perfection.

This is the classical specific definition of *Sat*, the Reality: It is always present in all the three periods of time, past, present and future and It is never subject to any change, It is unalterable, unchangeable. If you apply this test to anything in this world, you will find that all the things fail to fulfil this requirement. All the things change right from the beginning. Your body changes right from the birth, so it cannot be the Reality. Your human personality did not exist before you were born. The duration of its existence here on the earth plane is for a limited time. One day, everyone who is born has to go. Everyone knows this. It does not require any study of any big philosophy.

Death the Leveller: Gurudev used to sing: *Time sweeps away kings and barons.* Even the big people who wield great deal of power, were powerless before old age, disease, decay and death. The glories of our personality are shadows only, they are not substantial things. There is no power against fate. Death lays its icy hands on kings and barons. Even the biggest sceptre and crown must tumble down in the dust one day. An emperor may be holding a sceptre as a symbol of royal power and may be wearing a crown as a symbol of the royal status, but all these must go in dust and become as worthless as the dust.

The title of Gurudev's poem is "Death the Leveller." Death makes everyone equal, whether he is a beggar on the street-side or a great king, a president of a big state or an ordinary labourer breaking stones on the road-side or cutting wood in a forest. When death comes they all become the same, either they turn into a little ash or become food for the worms under the earth! This is a common human experience.

So Gurudev says:

Time sweeps away kings and barons!
Where is Yudhishthira? Where is Ashoka?
Where is Shivaji? Where is Napoleon?
Where is Valmiki? Where is Shakespeare?

Where are those mighty people? Where are Yudhishthira and Ashoka? One is the greatest from the point of view of

Dharma, the righteousness, and the other is great from the point of view of vastness of the empire. The empire of Ashoka stretched from South India and went beyond the North India and covered Pakistan, Afghanistan, Kabul and Gandhar. Archaeologists find his coins and Ashoka's inscription carved on rocks and stone pillars and in all these areas. All the great warriors, great ppets have disappeared.

Lord Krishna who is the creator, preserver and dissolver of countless billions of universes, who is the supreme Lord Himself, He opens the wisdom teachings of Srimad-Bhagavad-Gita and right in the beginning of His teachings says; "*Jatasya hi dhruvo mrityuh.* [Certain is the death for the born]." For one who is born, death is certain. That is the only certain thing, other things may be uncertain. Therefore, this human personality which did not have any existence before we were born, will come to an end one day. One may live very long, say a hundred years, hundred and ten years, but ultimately this life has to end. If it does not end, you yourself will pray to God: 'O God! Please take me away from here. It is painful to live here; I cannot stand, I cannot even sit, I cannot eat. Every moment is painful. Please relieve me from the burden of this body.'

Where do you go after death? The human personality is not there. That means the human personality is *asat*. Anything which is changing is *asat*. You were something in infancy, something different in your childhood, adolescent, youth and old age. Everyday there is a change in your mood, change in your attitude, change in your health condition. Everything changes. Instability is the second name of this body. This body is changeful, it has a beginning and an end. It is temporary in time. Therefore, it cannot be the Reality.

Your Essential Nature: Which is the unreal aspect of your being? What is *anitya, asat,* non-real and changeful? The Reality is always present, always unchanging; It is without a beginning and an end. Why? Because It is a part of the Cosmic Principle that is eternal, infinite, unchanging, permanent, imperishable. That Cosmic Being is called the Universal Soul or Paramatma. They call It God, they call It Allah, they call

It Almighty Father in the Heaven, or Ahura Mazda or Tao or the Supreme Reality. They may call It by any number of different names, but It is the same One, the same Universal Essence who is the source, the support and the fulfilment of all the existence. That is Perfect, That is Full, That is Divinity. You are a part of That in your inner Reality, in your essential Nature.

That Divinity is inherent in you. It is your true identity, essential identity, eternal identity. That eternal identity is immortal Spirit, That eternal identity is perfectly Divine.

This Reality is the very central Principle on which Gurudev has focused his message. He says: "You are immortal Atman; you are perfectly Divine. This life is a rare gift, a golden chance, a unique opportunity given to you. You should always make use of this life to become awakened within. Unfold this slumbering Divinity. Your endeavour should be to transform yourself into a Divine personality, into a Perfect Being."

You are not just a human personality with all the drawbacks and weaknesses; you are simultaneously a Divine personality. You are the Divinity on the earth. "Awaken the sleeping God within you," says Swami Vivekananda. This is the great task of your life. Live your life in such a way, make your thinking, feeling and actions in such a way, that all these become a great help for bringing about an inner process in this direction; so that they become a wonderful support for the specific techniques and practices for bringing about this unfoldment.

Gurudev gave his message as a message of Divine Life, living the life in such a way that the Divinity within you becomes awakened through Bhakti-Yoga, Jnana-Yoga, Raja-Yoga and various other Yoga systems and constant practice. If your inner life and spiritual practices are supported by a helpful, and become supplementary and complementary to your outer life which also partakes of the same quality of divinity, then your whole life becomes a unified process of trying to attain one single objective of the Divine Perfection.

That is the great objective with which Gurudev founded this institution; and to initiate you into the ways and means of bringing about this more rapidly, more effectively, he set up this Yoga Vedanta Forest Academy. His objective was to initiate you into the inner skills, the secrets of *sadhana* and practical spiritual life so that you can live actively an inner spiritual life.

The Vedanta Doctrines: All the Upanishads taken together are called Vedanta, because they comprise the Jnana-kanda, the end portion of the Vedas. Vedanta literally means the end portion of the Vedas. These Upanishads are the culminating portion of Vedas; it is pure *jnana*. So Vedanta means *jnana*. The Jnana-kanda of the Vedas is the record of transcendental Experiences and the ways and means of attaining It. So the contents of the Upanishads is about the nature of that transcendental Experience and about what happens to the person when he attains that Experience.

How many kinds of Vedanta are there? Innumerable. Because, unfortunately the learned scholars who study the Upanishads deeply, reflect over them and try to understand them, have understood them in various ways. So each one gives his own interpretation of what he thinks is the correct interpretation of the Upanishads. According to the interpretation given by that particular person who understood it in a particular way, a school of thought of Vedanta arises.

All the human beings are not the same. They have different intellects and different ways of understanding things, different approaches and different ways of looking at things, both physically, as well as, through the intellect. Later on, the scholars started to interpret the structure of the words of Vedanta. One school says: *'Tattvam-asi'*, the other tries to prove that it is *'Atattvam-asi'*. Due to such varied interpretations, certain well recognised systems have come into existence. The amazing part of it is that all these systems form valid parts of the total body of *Sanatana Dharma*. *Sanatana Dharma* does not take any side or validate any particular system. These are merely interpretations of the scholars and have become part of it and it tolerates them and treats as valid. So all these schools are called Vedanta.

The Absolute Monism of Sankaracharya (788-820) is called Kevala Advaita Vedanta Siddhanta, a little modification of which is known as Vishishta-Advaita Vedanta or Qualified Non-dualism of Ramanujacharya (11th century). Sankaracharya said *jiva* and *Brahma* are one. Ramanujacharya agrees with this oneness but interprets that it is so in a specific way only. In the Absolute Monism, *jiva*-hood is considered as a myth or an imaginary existence only; there is no real existence of *jiva*. *Ekameva-advitiyam Brahma, Neha naanaasti kinchit*. One and only One exists, there is not many. There is only One without a second.

He is uncompromising about this and as though to emphasise it, he says: 'If you think anything else exists, it is not real but it is your *bhranti* only.' You may argue: 'How do you explain *jiva*?' he says: 'What do you mean? There is no *jiva* to be explained. *Jiva*-hood is a myth, it is non-existent, it is a non-entity.' What he was trying to prove was his own *Aparoksha-anubhuti*, his direct Experience of the Reality, his Advaitic Experience of non-dual Consciousness.

Nimbarkacharya (1162) gave another interpretation known as Dvaita-Advaita Vedanta Siddhanta, the Dual-Nondual Doctrine of Vedanta. He says: "It is inexplicable thing! There is *dvaita* as well as *Advaita*.

Then came Madhvacharya (1199) who propounded Dualism or the Dvaita Vedanta Siddhanta. Individual soul and the Supreme Soul are proved by him to be different and separate entities. They can never become one. In essence the individual soul may be like the Supreme Soul, but it is a *jiva's* status, can never be the same, it can never have the same power or glory as that of the Supreme. He is Supreme, individual is under Him. The distinction, the *bheda*, is eternal and so they remain two, as *jivatma* and *Paramatma*.

Then came another Acharya called Vallabhacharya (1479) who gave Shuddha-Advaita Siddhanta, the Pure Non-dualism, which is also known as the Pushti Marga.

Five hundred years after Sankaracharya, another great scholar, great grammarian and a brilliant intellectual took his birth in Mayapuri, Nabadweep. He is known as Sri Krishna Chaitanya (1486). He expounded a certain school of thought of philosophy and mysticism, known as Achintya Bheda-Abheda Doctrine. He says:

'The nature of the absolute Reality is impossible for us to grasp. Our mind is a petty, finite instrument. How can it comprehend That which is supernal and infinite? How can it grasp that Being which existed before the mind had come into existence? How can the mind go to that Realm, because it did not exist when That alone existed? That is beyond the mind, It transcends the mind; the mind cannot reach That. The mind cannot even think of That. Whenever the mind thinks that it is thinking about That, it is only a thought or an imagination; it is not That. That Being is *achintya*, unthinkable. So the relationship between That and Its creation, including we all, the *jivatma*, cannot be thought about; it is both, oneness, as well as, *bheda,* the difference.

When a river ultimately merges with the ocean, you cannot distinguish the river from the ocean. There is no more any river, there is only the ocean. But you cannot say that the river water does not exist at all. It is continuously flowing and the river water forms a part of the ocean water, no matter even if it has got mixed up with it. The river water will be different from the water of the ocean at the centre. The river water cannot reach there. So long as the river has lost its separate existence, there is oneness or non-dualism between the river and the ocean. But so long as we cannot completely deny or entirely ignore the existence of the river water, there is *bheda,* a difference, there is duality.'

All these doctrines are having the same origin, the Upanishads or the Jnana-kanda of the Vedas. So all of them are known as Vedanta Doctrines. All these Acharyas were highly learned scholars. Each one of them gave his own interpretation of the Vedanta and proved his system, his view point, to be valid, using arguments, logic and all that. Our great people are so large-hearted that they recognised all these systems to be valid part of our Satya Sanatana Vedic Dharma. All of them are given a place in it and we study all of them. What is the height of generosity is that even Charvaka Doctrine is studied! It is a school of philosophy that does not at all recognise the existence of God. Ours is a great democracy of spiritual thought with complete freedom of speech and expression. The highest spiritual Experience is interpreted by different sages from different points of view, ability and depth. So have come different doctrines.

Veda accepts such divergence by saying; *Ekam Sad-vipra bahudha vadanti.* [There is only one Truth which the scholars describe differently.] It is also said: *Shrutir-vibhinnaa smritayopi vibhinnaa; tathaa muninaam matayo vibhinnaa.* [Vedas are many, other Scriptures are many; so also the views of the sages.] The views of scriptures also vary; there are different views. So also the opinions of sages and saints are different. This is to be expected in human beings as each human being is different. In Hindi it is said: 'All the five fingers are different.' Similarly, the siblings of the same parents are all having different temperaments, different nature. But the parents accept them all. Similarly, all the different doctrines are accepted by the parental Sanatana Dharma. The same one Being is termed differently, viewed differently by the sages.

There is one common Golden Thread in all these various schools of Vedanta. Whatever may be the nature of your relationship with the eternal divine Reality, one thing is certain. And all of them recognise that: '*You are Divine; and you ale inseparably related to the supreme Being*. That is the origin from which you have come; you are of That; and within you there is something of That.' All the Doctrines agree about this; no one disputes this. There is no disagreement; there is no difference of opinion.

This is the most important thing and Gurudev focused hundred per cent on this. This one thing is enough, you need not worry about other things. Let the scholars quarrel and disagree with each other; let them prove one thing and disprove another. You may not bother about all those things if you remember this one thing that *you are Divine.* Beyond your passing, temporary, apparent, imperfect human personality which is non-real according to the Vedantic Doctrines, there is hidden within you the unchanging, eternal Reality, Divine Perfection, All-auspiciousness, All-blessedness. It is the Beauty of beauties, incomparable and unimaginable. It is full of light, full of purity, full of everything. It is glorious and grand, radiant and shining. *That is you*. That is your real nature.

Awaken Your Divinity: When you are That, why should you be content in making your life merely a process of petty, little expressions of imperfect qualities of your mind, intellect and all that? When you have the potential, when you have the capacity to become glorious, perfect, pure radiant Being, why do you not opt for That? Why should you remain content at a much lower level? Why should you remain sometimes loving, sometimes hating, sometimes liking, sometimes disliking, sometimes very friendly, sometimes quarreling and fighting? This is foolishness. So our sages say: 'Come! Come! Wake up from this slumber of imperfect life. Awaken yourself to your Divinity. Unfold your Divinity. Attain that All-perfection. Do not opt for anything but the highest and the best. Do all that is necessary to attain your real Nature. Do not compromise.'

That is the ideal of Divine Life. That is the fulfilment of the fundamental discovery of the Truth within as expounded by the Vedanta. The discovery and declaration of the Vedanta about Divinity within you is: 'Within each one of you is the Divinity.' These are the words prompted by Gurudev. This is the declaration coming from the Vedic era. When the ancient sages discovered It, they declared this and they have made a wonderful device to preserve this highest wealth of the humanity, by setting up a system of unbroken succession of a lineage of master and disciple, of *Guru-shishya parampara*.

This invaluable treasure should not be lost. So a disciple should make him fit and worthy of that, so that the wisdom of the scriptures may be imparted to him. He can thus be inspired by the Guru to attain It during this very life. If he chooses ten such worthy disciples and imparts them the wisdom of the scriptures and even if two or even one disciple experiences It, it is enough. Even if we have a single candle lit, we can have from it another, from it another, from it yet another, and the chain continues and the Light will perpetuate perennially. This is the wonderful device of an unbroken lineage of master and disciple. This *Guru-shishya parampara* has succeeded marvellously and most effectively in preserving this great flame of living spirituality from the ancient times till now, so that even now, towards the end of the twentieth century, you are in a position to be told:

"You are Divine. Divinity is your birth-right. Divinity is your true identity. Divinity is your essential, true, unchangeable, eternal, real nature. That is your super, greater and real personality. Human personality is a subordinate personality. You have a dual personality; the Divine and the humane. In spite of all the drawbacks, defects and weaknesses of your human personality, you are still the Pure and Perfect; nothing can touch your this aspect; nothing can alter it. Even though, you are now subject to imperfections of your temporary, passing human personality, you are fully Perfect. Recognise this and rejoice. Tell yourself: 'I am always the radiant Divine within. I will not lose a single day, a single chance of trying to bring about the great inner awakening and unfoldment. In this very body, in this very birth, I must succeed in fully awakening my Divinity, experiencing That, becoming established in That.'

> *"That is the Goal, becoming liberated by attaining the full experience of your Divinity. That is the ideal of your life. That is to be attained right now, while living in this body; not in some unknown state after death, after the body is buried or burnt."*

Divinity is Your Birth-right: Is this possible? Yes, certainly. The history of our great country shows that in every generation, there have been such great *jeevan-muktas*, the liberated ones, who say: 'I have attained It. Why don't you attain It? It is your birth-right. You have been given this human birth to attain It, that is the purpose of your life.'

The human birth throws wide open the portal to this blessed state of Divine Consciousness, the Cosmic Consciousness. It is only the human personality which has power to think, reason, reflect, ask, probe, analyse, and discriminate. The very fact that you have been endowed with this rare gift of human personality is sufficient for you to fix the goal of attaining the fullest Blessedness; and that should always be your endeavour.

This is the declaration of the Vedanta: *Uttishthata jaagrata, praapya varaannibodhata*. [Arise! Awake! having reached the great (teachers) attain that Atman. Katha Upanishad 1-3-14.] Become awakened! Approach knowledgeable persons; they are there waiting for you. Come and ask, they are waiting for you to give the highest Knowledge. They expect nothing from you; they have got everything. They are Full and this Fullness spontaneously enriches the whole world; just as the spring season without any reason, spontaneously brings about abundance of flowers and fruits and turns the earth into beautiful greenery, makes all the lives to flourish. The saints are like that spring. So *Jagad-Guru Adi* Sankaracharya says: 'They are there; go to them. Become fully awakened as they have become. Attain Bodhi, the full awakening.'

That is what you are. You are fully awakened but you are in the sleep state and you are dreaming. You are *"Nitya shuddha, nitya Buddha." Buddha* means awakened. This is the eternal call of Vedanta. This is the Vedanta in nutshell for you. You must respond to this great call of Vedanta. You should never forget it. Vedanta is not to be stored in the books and lectures. Vedanta is to be actively lived for awakening. You must respond to the great call *'Uttishthata jaagrata, praapya varaannibodhata'* and say: 'Yes, I shall now wake up and attain Illumination. I am prepared to do anything for It.' This should become inseparable part of very breath of your life, every cell of your life.

Sublimation of Mind

You may open the door. We are keeping the door closed to prevent the monkeys. But when so many people are sitting there is no danger of the monkeys coming inside. Each human individual is more than ten monkeys. He has more potential for mischief than that of ten monkeys. Its potential is limited within its little range of instincts and given capacity. There is a humorous saying: *Brahmachari shata-markatah* [A *brahmachari* is equal to a hundred monkeys]. In old days, after being invested with the sacred thread, the boy was known as a *brahmachari* and was sent to a *gurukula*. There were no outlets like radio, television, cinema or sports. So the *brahmacharis* were either studying or making mischief.

What is the nature of mind? It is full of desires and strong compulsive urges. It is very uncontrollable. How much uncontrollable? It is said, it is as much restless, fickle and fidgety as a monkey. You know, the monkey is considered to be the most restless creature. Similarly, the human mind cannot remain steady on one object at a time even for a single moment. It constantly jumps from one object to another like a monkey. But not an ordinary monkey. Suppose a monkey has been drunk and it becomes intoxicated. Now suppose it becomes possessed by a spirit.

How much uncontrollable it will now become! As if to add fuel to fire, suppose a scorpion stings this monkey. If a scorpion stings someone, the agony is so unbearable that he will be screaming and shouting, jumping and dancing, holding this thing and that thing. When all these things are added together, can you imagine what will be the plight of this monkey, what will be its antics? They say, the mind filled with desires and without any discrimination is like this monkey.

Can Mind be Controlled?: No wonder, Arjuna expressed his doubt, when Lord Krishna went on describing very nicely, how to control the mind, saying: 'You may go to a lonely place where there is no disturbance, where the environment is good and quiet; in a clean spot, prepare a seat that is neither too high nor too low, made of cloth, a skin and *kusha* grass, one over the other; make the mind one-pointed, with the actions of the mind and the senses controlled, fix the mind on Me and meditate, make the mind steady as a lamp placed in a windless spot which does not flicker...' All this is very nicely described.

But Arjuna had known the havoc his mind had created a little while ago as described in the Chapter I, and here Krishna says: 'You do this thing and that thing and control your mind!' Arjuna says: 'Whatever you have said is good, systematic and scientific. But the only difficulty is it is not just difficult but it is impossible to control the mind.

Chanchalam hi manah Krishna pramaathi balavad dridham;
Tasyaaham nigraham manye vaayoriva sudushkaram.

[The mind verily is restless, turbulent, strong and unyielding, O Krishna! I deem it as difficult to control it as to control the wind. VI-34.'] Mind is more uncontrollable and fickle and difficult to grasp than the blowing wind. Even if we may be able to catch the blowing wind and make it steady, it may not be possible to control the mind. So Arjuna asks Krishna: 'What is the use of describing all this when it is not possible to control the mind?'

Without contradicting Arjuna, Lord Krishna says:

Asamshayam mahaabaho mano durnigraham chalam;
Abhyaasena tu Kaunteya vairaagyena cha grihyate.

[Undoubtedly, O mighty-armed Arjuna! The mind is difficult to control and restless; but by practice and by dispassion it may be restrained. VI-35] I quite agree with you. The mind is extremely difficult to control. But it is not impossible to control it. It is possible to control the mind."

Everything is done through the mind. Mind is the source of all the problems. The solution always lies where the problem lies. If you have a heart problem the heart is to be treated. If you treat the stomach, the problem can never be solved.

The whole problem of an individual soul engaged in the journey of the earth plane is not outside; it is inside, it is the mind. For example, people who have been born in a family which has been non-vegetarian for generations, for several centuries, are used to that type of food only. Now suddenly, one such fellow, joins a society of strict vegetarians. He tries to stick to the rules of this society. But every time he sees all his people taking non-vegetarian food, there is a problem. His mind urges him: 'What a foolish idea you stick to!' It becomes a temptation that is difficult for him to resist; and there is a tussle in his mind.

On the other hand, think of a person who is born and brought up in a family which is strictly vegetarian for centuries.

All the relatives and friends are vegetarians. In his body cell, and hereditarily, there is no inclination, no taste, no temptation towards the non-vegetarian food. On the contrary there is some sort of aversion to non-vegetarian food. For such a person there is absolutely no temptation for that kind of food. Even when he goes to a grand party where good many tasty, choicest delicacies are served, he asks for boiled potatoes, toast and salad only, and he is not the least tempted by the tasty non-vegetarian food.

Suppose, in a family everyone—grandfather, father, mother, elder brothers, are drinking and there always is a bottle on the table. The child daily sees them drinking. When all are away and he is alone, he also drinks out of curiosity. In the beginning he does not like it, but soon becomes addict to it. Now when he grows up, he is always tempted to take a drink whenever he sees a bar or goes to a party where drinks are served or when he is in a company of friends who are addict.

On the other hand a person coming from a family of teetotallers is never tempted by the sight of a bottle or a bar, never enticed by the friends or party. Where is the temptation? Is it in the bottle or the sight of bar or in the friends or the party? Certainly not. All these things have no effect on the second person. The temptation is not outside, it is in the mind only. The vegetarian person or the teetotaller has absolutely no thought in that direction. It is the mental condition of the person which makes all the difference.

Mind, the Cause of Bondage: In the same way, the bondage, the *Maya,* the *prapancha* is not due to outside objects. You become bound only because you have a craving for this object and because you have attachment or aversion for this person or that, you have likes and dislikes.

It is because you give so much importance to the objects and persons that you become bound. If you are not the least affected, influenced or moved by anything, then you will not be caught in the net of *Maya* even while living in the midst of all the worldly objects and persons. So the entire problem is subjective, the bondage of *Maya* or *samsaara* is subjective.

The same *samsaara* exists for all; but a child or a sage or a realised soul is not affected by it at all. The problem is within you only.

What is that problem? Why this problem does not arise for a cow or a dog or a bird? Because they do not have a thinking mind. They don't have desires and cravings, ambitions and expectations, schemes and plans for the future. Their problem is confined to a very limited, small area of survival. All that a non-human creature wants is what it can eat, what it can drink, where it can take shelter, how it can protect itself and is interested in satisfying its basic impulses and instincts like procreation. The instinct of fear is put in it, by the Prakriti, the Nature, as a safeguard, so that it will not rush to its own self-destruction. It is because of this fear instinct that the moment a fire starts in a forest, all the animals, birds and even small creatures run for safety, away from the fire, in the opposite direction. This is the Nature's device or mechanism of survival.

Samskaras, the Root-Cause of the Problem: No other being except man has the problem because only man has a thinking mind. He has the ability to interpret a meaning out of every sense perception. Everything a man sees, hears, touches, tastes or smells evokes different kinds of psychological and emotional reaction in the fields of desires and sentiments. Had there been no such reaction, there would not have been any problem. But he does evoke reactions, and the problem begins. A man is in bondage only because there are desires in the mind, because there are cravings in the mind, because there are attachments in the mind, because the mind is never content. The main reason for this is our *samskaras,* the mechanism of impressions of such sense experiences, such sense enjoyments in the past, and the *vasanas,* the propensity to such experiences and enjoyments, and inclination for repeating them in the present and in the future.

Samskaras are the in-built impressions in the mind like the grooves of different sounds on a gramophone record. What is its nature? What happens if you put your thumb on an ink-pad and then press it on a paper? Exactly the same impression has been reproduced and will remain intact even after a thousand years. But it is inert; it does not reproduce by itself more impressions and has no other effect.

But the impressions in the mind are different; they are not inert, they are living impressions, they remain alive in the *chitta,* the mind-stuff. They have an effect, they exercise pressure for repeating the experience. They are living impressions and have the potential of generating desires. They create a desire for exactly the same sense experience as one had it previously. There may be *shubha-samskaras* which are good for you or *ashubha-samskaras* which are not good for you.

When you play a gramophone record it reproduces the song exactly as it was recorded. But it cannot produce the voice that produced it. It only plays back. But what is the nature of the *samskara?* When a particular sense-experience is repeated over a prolonged period, it not only creates a latent dormant tendency and a desire in that direction, but that tendency becomes a *vasana. Vasana* may be *shubha-vasanas* or *ashubha-vasanas,* auspicious and positive or inauspicious and negative.

The *samskaras* are living and dynamic impressions. So when you are put in a similar type of situation, they become active and powerful and you are impelled to go after the same sense experience. They are so circumventing to temptations that they compel you to that experience. They are also called seed impressions as they are like seeds produced by previous experience and are capable to reproduce that experience like a seed producing a tree similar to the one from which it has originated. So long a seed is lying in isolation, in an unfavourable ground, it remains inert.

But if you give it a favourable ground beneath the fertile earth and water it regularly, it starts growing; it becomes a little seedling and then comes out a sprout with a few leaves also; and if the favourable circumstances continues it ultimately becomes a big tree similar to that from which that seed was produced. The new tree produces more seeds and every seed has a potential to produce the same type of tree again and again. Such continuous chain reaction is the nature of *samskara.*

The human mind, with its imagination, has potential of completely reproducing the same condition or the psychological state of being. This impels the mind to go and indulge in that sense experience and enjoyment. It is because of these *samskaras* and *vasanas* that a man is propelled towards the sense objects, get attached to them, craves for them, wants to indulge in them, possess them, keep them, snatch them away from others.

The mind has four-fold aspects:

- In the form of reasoning intellect,
- In the form of imagining, remembering and desiring mind,
- In the form of feeling mind having emotions and sentiments,
- In the form of the assertive ego. Once a desire arises, it is the ego principle which asserts: 'I want it, I will obtain it, I will enjoy it, I will do this, I will do that.' It now propels the body into activity. The root cause of all the actions is ego.

All these different modifications of the psychological level of the human nature are totally outward-going and make you a part and parcel of the outward world. So you are enslaved by the desires for the outer objects and forget your own real nature, your own real identity completely. You undergo all sort of emotions, elevation, depression, satisfaction, dissatisfaction, fulfilment, disappointment, etc. All these positive and negative experiences are due to your propelling yourself outside through four aspects of your interior psyche. This has created for you a big range of a net spread by your desires to catch the birds of sense pleasures.

Dva suparnaa sayujaa sakhaayaa
Samaanam vriksham parishasvajaate;
Tayoranyah pippalam svadvattyan-
Ashnann-anyo abhichaakasheeti.

[Two birds that are ever-associated and have similar names, reside on the same tree. Of these, one eats the fruit of divergent tastes, and the other looks on without eating. *Mundaka*

Upanishad 3-1-1.] Why the second bird is caught in the net of *karmas* and their fruits? Sankaracharya says, it is due to *avidya-kama-karma-vasana-aashraya-lingopadhi;* it is due to delusion, passion or desire, activities, craving on account of clinging to the adjunct of the body. So you become full of *raga, dvesha, klesha, kalaha,* passion, desire, anger, envy, jealousy, intense craving, dejection, despair, disappointment. One moment you are up, one moment you are down. Now you are constantly in the wide range of outward propelling desires and that keeps you in a state of flux and confusion. This causes the problem.

Be God-oriented: Yesterday we saw that all the things can be directed to that one and only Being, the sole Source. In That only can be obtained, lasting and permanent happiness, unalloyed, unmixed, total happiness, enduring, eternal happiness, happiness that would not decrease or change. That is Brahma, That is God, That is the Kingdom of heaven. Direct your entire being towards That. Direct your mind, intellect, emotions, sentiments towards That, make all your activities God-oriented by spiritualising them.

You may ask: 'All said and done, how can I implement it? Why should I do so? I am involved in this world. All my friends and associations are here. I have attachment for them, all sort of feelings for them. I am fully involved in my surroundings and the situations I face. This is a familiar ground for me. I am born into it and right from my infancy, it is a part and parcel of my life.

So it is spontaneous and natural for my mind, for myself, to move towards them. I had not to learn to do so. I belong to this world and this world is mine; so I naturally flow towards it. It is something most natural, something easiest to do for me. But now you are asking me to reverse that process and connect myself to that unknown Being, whom I do not know, who is absolutely unfamiliar and completely unknown entity for me. How can I ever do it? There is a funny proverb in Tamil: How can I know about the cremation ground until I die myself?

'How I can know That which is all perfect, bliss absolute, absolute unmixed peace and happiness? How can I think

> *about It? How can I meditate upon It? It is easiest for me to think of my mother or father, television or cassette-player, a cricket bat or ball because they are a part of my life. I have not to make any effort or undergo any training to think about them. But I don't know that unknown, unfamiliar Being. How can I have any love or affection, inclination or longing for That? How can I relate myself to It? I know my relationship with my brother or sister, mother or father. But how can I be related to That? When I don't know what It is, how can I meditate on It?'*

True, your problem is genuine. But the solution is quite simple: *Detach from here, attach There.*

> *'But how to do it is a big problem. It is quite easy to attach myself here, because everything here is familiar to me. But how to think of That?'*

So first of all, you should get yourself familiar, get yourself acquainted, to That whom you want to worship, love, adore, on whom you want to meditate upon, towards whom you want to move.

How you can get yourself thoroughly acquainted with That? There are three different ways. The first is the heredity. If you are born in a family where both the mother and father are devotees and they are doing *puja,* and *kirtan,* singing devotional songs, reading about the *lilas* of the Lord and the lives of saints, talking about Him, you will automatically develop love for Him.

Secondly, if you are not born in such a family, you should think about our rich heritage, you should get yourself acquainted with it by reading the spiritual books. There is a unique tradition in all the parts of India. There are some people whose sole occupation and allocation is to master the scriptures and to go to various places and narrate, expound and elucidate them. Many institutions make elaborate arrangements for such discourses regularly and provide the venue. They go to remote villages also and explain the scriptures in their own, simple, easy way supplementing it with illustrations from day-to-day rural life. This is a part of Indian way of life. So even an illiterate villager also knows some philosophy; he knows about Shabari, Ahalya and many such characters. He knows about *Bhakti*, he knows about God's grace.

Our scriptures are abound in *stotras* and *stutis*, hymns and paeans of the Lord, glorifying Him. Our ancient rishis have not missed, during the narration of the main theme, a single opportunity of introducing *stutis;* so we find them at frequent intervals. We develop great admiration and adoration, great reverence and veneration for the supreme Being. If you listen to the glories, grandeur and greatness of God and the desirability of that wonderful Experience of attaining that Being and attaining the supreme Bliss, Peace and Happiness, and going beyond all sorrow and suffering, it will create in you a longing for God, for that Experience and your life will certainly be changed.

Thirdly, once your taste is developed in this direction due to your family background and upbringing or through listening to the Scriptures, you develop a strong desire to acquire more knowledge and naturally you buy some books. Now the original texts of most of the scriptures are available and they have also been translated in all the Indian languages. So you begin to read and study our great scriptures in your leisure hours, prepare your own personal notes and discuss the finer points with like-minded friends. You slowly build up your own library of inspiring literature. What you argued to be unknown and unfamiliar becomes known and so close to your heart that you will spend most of your spare time on That and you don't like to waste your time on the worldly things. Now, you like to read about It, you like to think about It, you like to talk about It and you like the company of people with such tendency only. Now unknown That becomes your best friend and guide, the remote becomes the closest, He becomes a part of your life, you cannot live without Him.

Shravana-Sadhana

In all the paths of Yoga, shravana is considered as a very important *sadhana.* It is listening to the description, to a clear elaborate nature and the glorification of that Being, the transcendental Reality. In *Vedanta Vichara Marga* of *Jnana-Yoga* its importance is clearly accepted. In the Bhakti-sutras of Devarshi Narada and Maharshi Shandilya, a very systematic and scientific method of developing the devotion is outlined; and the very first step is *shravana.* If your temperament is not intellectual, if you are emotional by nature, if you have spontaneous ability to have affection, attachment and love for God, or if you are artistic by temperament, then you are naturally attracted to the *Bhakti-Marga.* You should develop deep and firm attachment towards your deity by the *shravana* of His greatness, grandeur and glories. In addition to whatever has been expounded in the Bhakti-sutras of Devarshi Narada and Maharshi Shandilya, the Maha-Purana of Srimad-Bhagavatam gives us a very systematic and practical method of developing devotion in gradual stages and progressing steadily on the spiritual path. Here also the very first stage is *Shravana.*

In the science of meditation of Raja-Yoga of Maharshi Patanjali, though it has not been specifically stated as a specific *sadhana,* it has been included in another hidden form. The second *anga,* the limb, of the *Ashtanga Yoga* is daily *svadhyaya.* In *svadhyaya, shravana* is more important way to grasp the right understanding. In ancient times, the main method of imbibing knowledge was listening to the teachings of Guru and other great sages. You may study at home, but you may have many doubts or you may not have grasped the proper meaning. So listening to others is necessary. The proper understanding of the scriptures and the basic truths of life is possible only after a complete purification of the mind and the heart, refinement of the intellect and deep faith.

In ancient times, an aspirant was approaching a guru with the request: 'Please elaborate the nature of the Reality to me so that I can meditate on It.' The guru used to say: 'First of all you prepare yourself for that higher knowledge. A gross mind cannot grasp subtle truths. First of all, the mind is to be refined and it is to be made pure and subtle.' Even if these truths are told to an impure mind, it will not retain their impressions. So *chitta-shuddhi,* the purification of the mind-stuff, purification of the interior and sharpening of the discriminating intelligence are the necessary pre-conditions. *Shravana* is of great help in this direction. So *shravana* is the preliminary *sadhana* that prepares you for higher *sadhana* and the higher knowledge of Vedanta. Why these preliminary *sadhanas?* If your mind is always filled with grossness and if you have no proper discrimination, it will always go after sense indulgence; and as long as the mind is craving for the sense pleasures, you will not develop a keen discrimination, your intelligence will never be refined. Our ancient *rishis* must have reflected upon this. So our scriptures clearly declare that: 'Have a clean discrimination and Realise it. Free yourself from the deceptive nature of the sense objects, free yourself from the harmful nature of sense indulgence, free yourself from the clutches of your gross nature by constantly exercising discrimination, everyday, every moment of your life.'

You must be a person of great discrimination, intellectual, ethical and spiritual discrimination, so that the material world cannot fool you, cannot trap you. Then only you will be really free and you will be able to imbibe the great wisdom teachings.

Know it well that there is nothing good for you in these sense objects which attract you, entrap you, enslave you. They are the root-cause of all your pain and diseases, sorrow and suffering, disturbance and restlessness. So long the mind is restless, there cannot be real peace or happiness. Whatever intense passion and desire you may have for the sense objects and the sense enjoyments and all that, you gradually get rid of that with firm determination and constant discrimination. You convert your *raga*, passion, for the sense objects into *viraga,* dispassion, and then develop *vairagyam,* detachment. You will have to exercise very strict discipline on your mind.

Because the senses have been habituated with certain behavioural pattern over a long period and you have never tried to curb them, you have never tried to exercise any *samyama,* sense-control due to your delusion and ignorance. You have never come in contact with that type of people, that type of a society, that type of the world that do not run after the sense pleasures, but are trying to curb them. So you think you must have sense enjoyment. Soon you are accustomed to have sense enjoyments. Now it won't leave you unless you make a very determined effort.

Develop that firm determination. You must control the turbulent behaviour of your mind. You must control its constant chasing of the sense pleasure. You must try to restrain yourself. This is one pattern of disciplining your mind.

Sadhan-Chatushtaya

In the ancient times, the *guru* used to tell to a new seeker: 'First of all, practise the four-fold *sadhana* of *viveka, vairagya, shat-sampatti* and *mumukshutva;* and then only, I will give you the knowledge of the subtle truths of philosophy and Vedanta. Otherwise, even if I tell you these subtle truths, you will not be able to understand the correct implications of what I teach you, you will not be able to retain it. This requires purity of the mind and subtle intelligence.' Therefore, first you get yourself established in four-fold *sadhana;* then only you are qualified to do *shravana sadhana* according to the Vedantic *Jnana-Marga.*

In *sadhan-chatushtaya,* the four-fold *sadhana, viveka* or discrimination is the first aspect, *vairagyam* or dispassion the second.

Shat-sampatti: The third *sadhan* is *Shat-sampatti.* It includes *shama, dama, uparati, titiksha, shraddha* and *samadhana.*

Shama: The determination to subdue these will come only when you realise: 'As long as I allow the senses to push towards the sense objects and allow myself to get caught in the excitement of the sense indulgence, my mind will always be restless, agitated and disturbed, it will be in tumult and it can never be calm or serene. I can never have peace and happiness.' Once the mind realises that this is not the proper direction, it will say: 'No I will not think of the sense objects. The senses naturally plunge in the sense objects, only if I think of them.

The problem arises in *my thinking* of the sense objects. The sense objects are inert and non-intelligent; and so they cannot trouble me by themselves. But I myself allow them to become troublesome; because I do not control my mind, because my mind is full of desires to pursue them, enjoy them, experience them. So long as I have desires for them, my mind dwells upon them. Now I shall not allow it to dwell upon them. I shall divert my mind on to higher things. I shall cultivate different interests, other than mere sense indulgence. In this way, I shall be able to wean my mind away from its habitual propensity of chasing the sense objects and dwelling upon them.'

Now the mind will say; 'Like a lotus remaining unaffected by water even though remaining in the water, I shall remain unaffected even while living in the midst of the sense objects. Even if they come in my gaze I will not look towards them, I will withdraw my attention from the sense objects. First of all I will withdraw the senses from the sense objects; but if in spite of that the objects come into my perception, and if the linking sense is aware of it, I will withdraw my connection with that sense; and if it is too late, I will refuge to identify myself with the sensation that has already been created. Even if an object has come in contact with the sense, the sense sends the message and there is a cognition of the sensation, I will refuse to identify myself with that sensation. 'In this way, I will exercise triple withdrawal: The first and foremost, the most effective and the best, is the withdrawal of sense from the sense object. Secondly, it is the withdrawal of the mind from the sense. If both these fail and the sensation already reaches the mind, I will refuge to identify myself with that, I will not become an enjoyer of it, but will remain merely a pure witness to the happening and will remain absolutely detached and unaffected.'

As the mind begins to practise constantly this inward, invisible, subjective discipline, gradually it becomes free from the influence of the sense objects, free from the outer world of perception; it does not move towards the sense objects but thinks of higher things. And the mind ultimately attains a state of quiescence, absolute serenity. This state of quiescence is called *shama. Shama* is a discipline of establishing a state of serene quiescence in the mind. The mind will no more be restless, agitating or outward-going. Such inner stability must be supported by disciplining the senses, curbing of the senses.

Dama: Dama is controlling of the tumultuous and extremely uncontrollable activity of the senses. Just as a rider of a chariot exercises a firm grip over the reins and controls the horses, exercising a constant vigil to check them and not allowing them to move in a wayward manner; in the same way, the five horses of the senses are checked and controlled by the awakened intellect endowed with keen, active discrimination. Now it does not allow them to sway and to do what they like. Now you are the master, you give proper directive, you determine the path on which the five horses will have to tread. Now you have the control and you do not allow them to do what they like. This is known as *dama,* the subduing of the senses. This is the first step in the third aspect.

Uparati: This three-fold discipline will bring a complete transformation in your entire nature. The natural tendency of the mind and the entire psyche is centrifugal, it is to propel out and chase the outer objects of attractions and sense pleasures. It moves in all the directions towards thousands of objects and the world is filled with countless objects. The bees are always wandering here and there, never sitting quietly. But if the queen-bee moves to another place, all other bees will abandon their chase and follow it. Similarly, when the mind settles on higher things, all the senses follow the leader and give up their natural tendency to chase the sense objects. If the leader of a political party decides to change the alliance, all the cadre of that party will automatically join that alliance and become a part of it.

Similarly, if the mind carefully observes and analyses the hollowness of the glitter of the attractive objects of this *Maya-bazaar* and of the repeated sense experiences, and starts discriminating and turns away from the sense objects, it will be awakened. It is like joining a new alliance by the leader and the senses will follow it. When the mind becomes established in that state, a peculiar marvellous thing happens: the senses give up their in-born natural propensity to go towards the sense objects of attraction and temptation. The senses which were the main cause of the problem, give up their fundamental nature and the problem automatically ceases to exist. The senses now follow the new nature of the awakened mind, the mind full of discrimination, dispassion and detachment, the mind which is desireless and free from passion. When that happens, it is known as the fourth *sampatti* and is called *uparati*.

Titiksha: The third factor that obsesses and troubles the mind is the unpleasant experiences which upset you immediately. Too much of heat or cold, some little discomfort or inconvenience upsets you easily. Be a Spartan and develop a habit of bearing these without being upset. Don't leave all your work to correct them and to restore that comfort immediately. Don't allow it to distract you from your work, from your path. As the saying goes; Grin and bear it. This is called *titiksha.* It means to bear the little discomforts and inconveniences, troubles and unpleasant experiences, little ups and downs, which are part and parcel of everyday life. If you go on being upset by them and try to correct the situation, your time and energy will be wasted and your psyche will be affected. You should bear them and ignore them unless they are of a very dangerous nature.

Shraddha: *Samadhana* is possible only if you have absolute faith in whatever ideal you may have chosen. You should convince yourself: I know I have chosen the right path. Many have treaded this path and attained the supreme Being. They say this is the greatest Experience. They cannot be wrong. They had no personal axe to grind. They had no thread attached to it. They have nothing to gain or lose whether we adopt it or not. They did not want us to suffer. They wanted us to attain our highest welfare. They were our well-wishers, they have motiveless goodwill for us. It will be foolish for me not to have complete faith in them. So I have total faith. This is the sixth *sampatti* known as *shraddha.*

Samadhana: When one is well established in that state, he resolves: 'That and That alone is my goal. One-pointedness will be my state. I will not be attracted by anything other than That. There will never be any lack of self-control. I will never allow my aspiration to tone down even a little. I will not stop till the great goal is attained.'

Now your daily routine is totally changed. You are no more interested in petty things; you are no more wasting your time and energy after them. All your energy is fully directed in the inner direction, towards your goal that you have accepted. This is the fifth *sampatti* called *samadhana,* giving oneself entirely to a single ideal that you have adopted.

Mumukshutva: The fourth *sadhana* is *mumukshutva*, a strong desire of final emancipation. If all the things described above are to be kept up at the same pitch of intensity, you must have a very keen and intense longing for attaining Liberation. If you become a little slack, if you allow your effort to be diluted a little, finished. You may fall back to the square one.

What a great thing, great fortune it was to become one of the twelve disciples of Lord Jesus Christ! But one of them was tempted a little by the glitter of the gold and betrayed his Master. This is symbolic. Even after reaching such a very high state, if you allow your keen, intense involvement in that state to be diluted a little, finished. Therefore, intense longing for the goal is necessary.

Jeevo Brahmaiva An-Aparah

Sri Jagadguru, Bhagavadpada, Adi Sankaracharya is one of the greatest philosophers India has produced in its long chain of great philosophers. He is also regarded as one of the greatest philosophers of the world by good many philosophers and thinkers of the whole world. He was a giant intellectual personality and no one has equalled him till now on the scene of philosophical literature of the world. His contribution is simply amazing. He is regarded as a phenomenon.

In a brief life span of thirty-two years, he has brought out voluminous philosophical, spiritual and devotional literature of such high standard that it has set for ever the norms for classical spiritual literature for the entire world.

For a long time his literature was confined to India only. But its richness, the towering pinnacle of its conclusions, sheer strength of its logic, its crystal clear rationality are so remarkable and irrefutable that once some German scholars read him, they immediately started translating his works into German language and soon his works attracted the attention of the European scholars. The German translations were soon translated into other European languages.

Maharshi Veda-Vyasa gave us the quintessence of the Upanishadic philosophy in Vedanta-Sutra which is also known as Brahmasutra and he was indirectly responsible for bringing to us the great spiritual discourse of Lord Krishna to Arjuna. This *jnanopadesha* is known as Srimad-Bhagavad-Gita. It covers all the aspects of spiritual life. As it is woven in the great epic Mahabharata written by Maharshi Vyasa, he is indirectly responsible for bringing to us the divine preaching of Bhagavan. He also gave to the world the elaborate and very much voluminous eighteen Puranas. After giving to the world such unthinkably vast literature, he has said something very peculiar, very extraordinary, towards the end of his life: 'I have brought out this Puranic literature with the welfare of the mankind at my heart. In all this voluminous literature I have brought, I have told you what is meritorious to human activity and what will lead you to your greatest good and also the opposite of it, what is demeritorious or what you should always avoid, because it leads to your greatest harm, to greatest sorrow. I will sum this up in half a verse:

Ashtaadasha-Puraneshu Vyasasya vachana-dvayam;
Paropakaaraaya punyaaya paapaaya para-peedanam.'

[Vyasa has to say only two things (as quintessence) of the eighteen Puranas: *to do good to others is worthy, to harm others is unworthy*]. To harm, harass or injure others, or to inflict pain on others, to cause sorrow to others, is the greatest demerit and one has to suffer for that. If you make your God-given gift of the human life in being good to others, utilise it in doing good to others, in benefiting life around you, to help all creatures, that will lead to your own highest good. It will result in your own happiness, in your own supreme welfare.

Jagadguru Sankaracharya, this towering philosopher has given the quintessence of his colossal, towering philosophical literature in a similar way. At the end of his career he says:

'Look here! Listen! All that I have written about, all that I have commented upon, I sum up in half-a-verse:

> *Shloka-ardhena pravakshyaami yad-uktam grantha-kotibhih;*
>
> *Brahma Satyam jagan-mithyaa jeevo Brahmaiva an-aparah.*

[In half-a-verse I shall express what has been said in crores of books: Brahma alone is real, the world is *mithyaa;* the Soul is Brahma and no other.] The Reality alone is. The Supreme Existence is eternal, non-dual and without a second. *You are That only.* Everything perceived through the five senses and understood by your mind and intellect is unreal. All the objects of the world are like the water of a desert mirage, they are like the glint of the nacre, the mother pearl lying on the sand of seashore and appearing like silver, *shuktikaa hi rajatavad-avabhasate.* They have no real existence but they have only seeming appearance and that too temporary.

What appears in a dream as cent per cent true, true beyond doubt so long as you remain in the dream state, disappears in a fraction of a second once you come to the waking state. Similarly, this world disappears once you are awakened. The dream experience is so graphic and so complete in all respects, and you perceive every detail, and you experience it through all the five sense-organs, that experience dominates your consciousness completely and during that period that experience remains the only truth for you.

But when you wake up suddenly due to the ringing of the alarm-clock or any such disturbance, the very next moment everything disappears, everything becomes unreal! You cannot say that your dream was non-existence or unreal because you remember it in full details; yet you cannot say it is real because you cannot photograph it, print it, or you cannot reproduce the conversation and the scene or the objects and the characters in any other way.

Nevertheless, one part of your consciousness retains its memory and you can write it down or narrate it. But now it is non-existent. The moment you wake up, it becomes non-existent and irrelevant. Now your consciousness is dominated by some other events of some other reality which takes over the charge of your consciousness.

Our seers say, it is just like that. You are a part of this dream experience, this dream world, and you think that it is real, your outer personality is real. But when you wake up to the supreme, cosmic Reality, your true identity, then only you know that your false identity was as much a figment of that dream as the characters and all the factors of that dream world. You thought, you knew yourself as Mr. So and So, you conceived your identity with a certain name and form; but when you wake up, when you attain the *Bodhi,* the Enlightenment, the Illumination, that very second you yourself become Light only, your false identity disappears completely, like the disappearance of darkness as soon as you put on the light. You may wonder: How could I cling to the darkness all along! Why was I thinking like that? What a foolish way of thinking!

Sankaracharya does not stop at that. He asserts that your present personality is false; because you are no other than that supreme, wonderful Being. You are one with It. In a most wondrous way he is able to tell us graphically how it is so, so that, with a little common sense one can easily understand it and it becomes absolutely clear and beyond doubt. The world we see and we know with all our sense perception is not real and non-eternal. It is there for a moment and ceases to exist the next moment. It is your delusion to think that you are a weak and imperfect, limited being confined to this physical body and the name and form associated with it. Remember! *You are no other than That, you are the Brahma-tattva only.* You are the cosmic Being, you are the Perfect, Divine, Universal Being. You are no other than That.

How is it possible? It is *svatah-siddhah,* self-evident. Because Brahma is *ekameva-advitiyam,* the one and the only Existence, without a second. There is no second. So your Soul is no other but That only.

Our seers have proved this in an amazing way, so that an ordinary man can also easily understand. In an ocean, there are innumerable waves; some small, some big, some very big, some giant size. They are of different shapes and sizes. You can see them, you can observe them dancing and playing. You can distinguish them, identify them. You can photograph them. You can hear them roaring. There are innumerable waves with different shapes, characteristics and different life spans. Each wave has its own separate identity, its own existence. Even when you are watching at these waves with different individual shapes, sizes and characteristics, what you are looking at? Do you see anything other than the ocean?

What is the origin of these waves? From where they have all emerged? Do they really have separate individuality? How the seeming individual existence of these waves was possible? When they broke away, where they went? Each wave arose from the ocean. It owed its existence to the ocean. When it momentarily existed as a seemingly separate identity, it was always supported by the ocean and always remained a part of the ocean. It never had an existence independent of the ocean. When it subsided, it merged back into the ocean only. Thus, ocean is its *aadi, madhya* and *anta,* the beginning, the middle and the end; that is, it emerges from the ocean, its existence is possible with the support of the ocean only, and when it subsides it merges back into the ocean and once again becomes one with the ocean. Even when it momentarily looked something distinct from the ocean with a separate individual identity, it was at no moment other than the ocean itself.

What is the implication of this? Like the wave being ever no other than the ocean, the individual soul is never other than That. The ocean is the only real existence; the apparent multiplicity in the form of waves has no ultimate reality. No other thing besides the ocean ever exists.

There is such a large variety of clothes with different colours and different designs. But beyond all these forms, what exists is only cotton *tattva;* nothing exists there other than cotton. If you go to a big pottery store, you find hundreds of different articles of different sizes and different names and different forms. Even when you are seeing these variegated types of pottery articles—delicate, big, small, artistic, decorative, utilitarian—you are looking at one thing only, the clay. All the pottery is made out of clay only. You may give different names to pots, but they are nothing other than clay. There are different ornaments, but a goldsmith sees nothing other than gold.

These are some of the classical analogies given by our ancient sages, in order to bring home the non-duality of *Brahma-Tattva*, in spite of so many variegated manifestations of that *Tattva* in different names and forms. There is only one Existence, beyond all these multifarious names and forms. While explaining this to a Vedanta student, the teacher does not elaborate, but simply says: *Ghata-pata-adi nyaayavat. Ghata* means pot and *pata* means cloth. So it means, like the analogies of pot, cloth, etceteras, there exists only one *Tattva,* even though seemingly there are so many, and are seen as many.

Some people give another analogy: A spider decides to make a home for itself in a corner of a room. If we have to build a home for us, we have to buy so many different things—bricks, cement, iron rods, various items of hardware and sanitary-ware—from so many different shops. A bird builds a nest as its temporary abode. It goes here and there and collects so many things from various places. From how many sources the spider has to obtain all the materials? What are the materials? It does not go anywhere to collect the materials. Whatever is necessary to make a web, it brings out of its own self and remains in that web; and when its purpose is served, it absorbs the web back into its own self. Can you ever say that the web is different and separate from the spider? In spite of there being seemingly two different things, the spider and the web, there is non-duality. The web is no other than the spider.

What is the purpose of the various philosophical systems and vast spiritual literature? Why do they exist? They exist to tell us certain home-truths. Why should they tell us these home-truths? What is the value of these home-truths? Because if you are not aware of these truths, you will be living in delusion, you will be deceiving yourself. You can live in wisdom only if you are fully aware of these truths; you will then no longer remain in delusion.

The second important, pragmatic necessity of knowing the truths is: The eternal quest of the entire mankind in the whole universe is: How to overcome pain and sorrow and obtain lasting happiness? Our scriptures have summed up the answer to this in an amazing manner. All the people are in quest of so many things. A farmer is engaged in trying to get a proper field, proper manure and fertilisers, proper seeds, proper pair of bullocks, proper equipment, water for irrigation, this and that—hundreds of different things.

Similarly, people in other occupations are also trying to find so many things. But the scriptures say: 'Ultimately, all are engaged in search of one thing only. That one thing is happiness. No one wants pain, sorrow, grief or suffering. Everyone wants to avoid unpleasant, undesirable, painful experiences. All want happiness and they want to perpetuate that experience of happiness. *Sukham bhuyaad, maa bhuyaad duhkham.* Avoiding sorrow and suffering and trying to obtain happiness.' This is the eternal quest of mankind in all the parts of the world. This sums up in nutshell, the eternal quest of the man on this earth. We may do it in so many different ways, but our objective is the same.

If you know certain truths of human life and base your life and conduct on them only, you will be able to get rid of sorrow, suffering and pain, and obtain real happiness. Real happiness is not momentary but permanent. If you do not base your life on these truths, your quest will be a leap in the dark, a wild-goose-chase. You cannot get permanent happiness from those things which are themselves temporary. And everything in this world is temporary.

So, the worldly things cannot give you happiness. Therefore, you should not crave for anything in this world; you should not go after anything in this world; you should not allow yourself to get enslaved by the temporary objects of this world. Everything in this world is temporary; there is not a single thing which is permanent. What are the adjectives used by Lord Krishna for this world? *Anityam, asukham, ashashvata,* impermanent, unhappy, non-eternal—these are the adjectives with which the Lord refers to His own creation, His own work. Our experience teaches us that even while some happiness seems to be present sometimes, it is not eternal, it is for a short duration only; and you cannot rely on it because that experience of happiness is always changing. One moment it is there, the next moment, before you can grasp it, it disappears or diminishes.

Why do you crave for it? Your imagination of a thing may be something and you may think: 'I will be happy if I get it.' But when you get it you find it to be totally different from what you had imagined. Now you want to get rid of it! You are imagining something, but the actuality is something else. And even if the actuality tallies and coincides with your imagination, it goes on changing while you have it.

Here is a simple analogy. You want to enjoy your Sunday with an ice-cream party. The carry-home service delivers you the best quality ice-cream. It is served in beautiful bowls to each member of the family. It is decorated with cherry and special cream. Everything is ready. But the telephone call comes and you have to attend it. It is some urgent and serious matter. You ask other people to take the ice-cream and you continue to talk on the phone for twenty minutes. When you come to the party again, where is your beautifully decorated ice-cream? It is no longer what it was. It is neither cold nor does it stand, it has already dissolved. So also, all the things in this world are changing. Even the most beautiful or the nicest thing will also become worthless.

Suppose, you have brought the best quality alphanso mangoes. As soon as you enter your home, there is a phone call from your boss. He wants you to rush immediately to some place for a few days on some urgent important work. You leave home in a hurry, forgetting to tell your wife about the mangoes. When you come back home and remember about the mangoes and open the packet, what you find is that all the mangoes are spoilt and some juice is coming out of it. You had expected the fruits to give a great joy; but now you have to do the cleaning and throw them away. Is there any intrinsic lasting value in the mangoes? All the objects of this world are non-eternal and yet we are always clinging to them!

Suppose, a boy meets a very beautiful girl. He is so much enchanted and enamoured that he thinks he will never be happy unless he marries that girl. They got married. After some time, the girl is severely infected by small-pox and it makes her very ugly. The love of the boy changes completely. He says; 'Where is that beautiful girl whom I had married. She is not that beautiful dream-princess of mine.' So Gurudev says, what attracts the mind is the outer physical beauty.

Suppose, you like *khir* very much. You are invited by your bosom friend for a party on some festival day and the menu includes *khir*. Its preparation is most delicious. You eat beyond your capacity. Now your friend fondly insists for a couple of helpings more and you take some more *khir*. Now it is the turn of his wife. You cannot say no and the *khir* causes you discomfort and nausea. The same *khir* which you like very much is now the cause of all the trouble. Where has gone that pleasure-giving power of the *khir?* If it has in itself an intrinsic power to give you happiness, your happiness should increase more, the more you take it. But on the contrary it causes you discomfort. That means the experience of happiness is subjective and is not in the outside object; the object has no inherent quality to make a man happy.

All these analogies bring out one common truth: There is no happiness in the objects of the world. They are the cause of pain. Lord Krishna says:

Ye hi samsparshajaa bhogaa duhkhayonaya eva te;
Aadyantavantah Kaunteya an teshu ramate budhah.

[The enjoyments that are born of contacts are only generators of pain, for they have a beginning and an end, O Arjuna! The wise do not rejoice in them. B.G. V-22]

The first home truth is: All want happiness. Why do you then chase those outside objects which have no intrinsic capacity to give you happiness? If you want happiness, make effort in the right direction. Time fleets away. You are here for a limited period only. The rare chance of human birth is not endless. Therefore, be up and doing. If you seek happiness, exert yourself in the right direction.

What is the wrong direction? The world is the wrong direction; anything and everything in this world is the wrong direction. There is not even an iota of happiness in the objects of this world and they can never make you happy. This creation of Brahma has some utilitarian value; but it has a nuisance value as well. It can never give you happiness.

The second home truth is: The real Happiness is in That All-perfect, non-dual *Brahma-tattva,* One without a second. All the objects of the world are merely generators of pain, sorrow and suffering. Therefore, don't waste this precious opportunity by running in the wrong direction and ultimately getting caught in disappointment, frustration, sorrow and lamentation.

Which is the direction that you should follow? You need not study deeply the scriptures and big volumes of philosophy to learn this. Many saints, mystics and spiritual seekers have treaded the right path and reached their destination and attained That. They implicitly followed the right direction, obtained that experience and obtained the Happiness. Whoever has followed them had also the same wondrous Experience. Make a deep study of their lives and try to follow their footsteps. A poet says:

The lives of great men oft remind us
That we can make our lives sublime.

We should emulate the lives of the mystics and saints who have attained That; we should follow their footsteps. They were people who lived a very austere life, they were men of *vichara* and *viveka;* they never led a haphazard life.

Therefore, always do *vichara* and have *viveka,* enquire and discriminate every moment, every step, every breath of your life. Let there always be active discrimination and enquiry within you. You have not to worry now, because you will not miss your step and go in the wrong direction.

The third home truth is: It is the lives of the saints that teach us in what direction we should go, how to lead our lives and what practices to engage in, so as to attain the highest state of supreme Blessedness and Bliss, of eternal Happiness beyond sorrow and suffering. *Mahajanah gataah sa panthah.* That is the path which the great ones have trodden. That is your path. That is the path you should follow.

When they attained the supreme state of Blessedness and Bliss, they found that That Blessedness and Bliss was no other than the eternal *ekameva-advitiyam Brahma-Tattva*. The very nature of that Supreme Reality is the quintessence of one hundred per cent, unalloyed, pure, total Bliss. So they declared: *Anando Brahmeti vyajanaat.* [He knew Bliss as Brahma. Tait. Up. 3-6-1.]

Therefore, strive by all the means possible, through thought, word and deed, through whatever action you may engage yourself, to connect yourself to that supreme Reality. Because from that with which you connect yourself, its nature will flow into you. This electric bulb is connected with the electricity and so it gives you the light. But if you cut off this connection by switching it off, there will be darkness; because now the nature of the electricity is not flowing into the bulb. So, why don't you connect yourself with That which is Light of lights, which is Bliss-absolute, Peace-absolute, which is eternal Happiness? *Teshaam shantih shaashvati netareshaam.* [To them belongs the eternal Peace, not to others. Katha. Up. 2-2-13.] So connect yourself with It. Link yourself with That. That is the essence of all the *sadhana,* all the philosophy. Establish contact with that supreme, transcendental Reality. Make a permanent inner link-up with That and go on striving for It by every means possible and move in the direction leading to It. Go on increasing that relation by leaps and bounds and make your progress in geometric progression. Have full faith and you will reach That.

That is the essence of all Vedanta, essence of all Yoga, essence of the *sadhana.* Connect yourself with That, sustain that connection and move towards That more and more. Intensify your relationship with It. Make it more intense, and more and more intense. That is the purpose of all *upasana,* that is the purpose of *japa,* that is the purpose of all *kirtan,* that is the meaning of *aaraadhanaa,* that is the meaning of meditation. All these are the means or processes of building up and consolidating that relationship, the spiritual link-up with the Divine. All these are to move progressively towards the supreme Reality, till you ultimately attain Oneness with That.

When a small stream comes out of a glacier of a snow-clad mountain, it is so narrow that you can jump over it and go to the other side. But slowly more streams join it, and its flow goes on increasing. It becomes a small river. Some other rivers join it, and it becomes a big river. There is so much water! But it goes on to its chosen path to meet the ocean non-stop, continuously, untiringly. If there is some obstacle in the form of a rock or hill, it remains undeterred and finds its right path again. Ultimately, it does meet the ocean and becomes one with it. That is the success of its life-long effort. That is the fulfilment of its life. Your life should flow like that continually on the same one path, progressively coming nearer and nearer towards that supreme Being. Stop not till the Goal is reached. This is the call of Upanishad: *Uttishthata jaagrata, praapya varaannibodhaṭa*. [Arise! Awake! Having reached the great (teachers) attain that Atman. Katha Upanishad 1-3-14.]

This is the quintessence of all the philosophy—eastern, western, ancient or modern. All the philosophies exist to tell you that the greatest Bliss is in That, the One, non-dual; and not in the many fleeting appearances. It also tells us that this human birth is a rare opportunity and so you should not postpone your march towards the Goal, you should not waste a single moment. Be up and doing. Stop not till the Goal is reached.

All the scriptures, all the philosophies exist only in order to liberate mankind from the web of sorrow in which he is caught, and to take him beyond all sorrow, pain and suffering; and to make him attain and become established forever in the state of supreme Bliss, Perfection, Blessedness and Liberation.

Hari Om Tat Sat.

(At the end of the discourse a student asked: "Is it possible to attain such a state in this very life?"

Sri Swamiji assured, *"Emphatically yes. It is possible, one hundred per cent."*)

Objective of Yoga-Vedanta

You are the students of the twenty-fourth Yoga-Vedanta course of the Yoga-Vedanta Forest Academy of this spiritual,

Dharmic and ethical institution founded by Guru Maharaj Sri Swami Sivanandaji sixty years ago. First, you should acquaint yourself with what you are engaged in.

The first objective of this course is: You want to know something about the Indian culture, about Yoga, Vedanta, meditation and about the Indian scriptures like the Gita, Upanishads, Bhagavatam, Ramayana. That is, you want to fulfil your *jijnasa,* your thirst for knowledge.

Secondly, in some of you there may be an urge: 'I am born in India. If I don't know anything about the Indian culture, I am an Indian in name only. If I have to become a true Indian, representing my culture I must absorb and assimilate the Indian culture so that I may live according to the ideals of our culture and these ideals may manifest in me. Then only, I can become a true Bharatiya.'

In addition to these, in some of you there may be a still higher third dimension of spiritual longing, the spiritual *jijnasa, mumukshutva,* a spiritual hankering: 'This course may help me to become a spiritual person. It may help me to put me on the path of spiritual seeking and spiritual progress. It may help me to move towards God.'

All these three objectives can be achieved here, if you live here not only as students but also as persons seeking spiritual unfoldment of your personality, if your approach is: 'I want to move ahead in the life, I want to make progress not only in the dimension of intellectual knowledge, not only as a lover of Indian culture, but also in the dimension of my spiritual Reality.' We are not merely human beings. This personality, this identity is merely a temporarily added token identity. You were not having this human body before you came to this earth plane, before you were born to your parents. You did not have a human name and form, a human personality or even the human consciousness. In that state you were something else; what it is you do not know.

If you think deeply about the life, you also realise that this temporary human personality shall end one day. One day you will have to discard this human identity.

The whole of our religious thought originates from a mass of Knowledge and Wisdom. It does not originate from any specific single, divinely inspired outstanding personality like Buddha, Mahavir, Zoroaster, Moses, Jesus, Mohammed or Guru Nanak. There is no such personality behind the Satya Sanatana Vaidika Dharma. It is timeless. It is so ancient that its beginning is not known to anyone. The end portion of the Vedas, known as Upanishads, is the treasure of the mass of knowledge, wisdom arising out of the highest supernal Experience. The earlier portions of the Vedas may be still more ancient which goes beyond the recorded human history. This is the only non-prophetic religion in the world.

There were some ancient civilisations like the Assyrian civilisation, the Egyptian civilization, the Greek civilization, etc. Whatever way of worship and living they were having, were dubbed by the occidental scholars by the common label of paganism. Those scholars thought that the Hindus were superstitious and they always criticised them. But when they came to discover the towering heights of the philosophical wisdom that we have attained, they were bewildered, they were stunned. They started looking at our religion with reverence. Their scholars started studying the Vedas. One of the first people to study the Indian scriptures were the Germans. They found in the Vedic wisdom profound depth. A great scholar Max Mueller translated the Vedas in the German language. The English translations are from the German translations of Max Mueller, Paul Deussen, Schopenhauer and other outstanding German philosophers who studied the Indian philosophy. Schopenhauer was so much inspired by the Upanishads that he said: "Upanishads have been the solace of my life and I know that they will be solace of my death also."

The quintessence of all the Vedas is in the end portion, the *summa cum laude,* the epitome is the Upanishads; and the essence of Upanishads is the Bhagavad Gita. Bhagavad Gita starts by telling this terrible truth: *Jaatasya hi dhruvo mrityuh* [Certain is death for the born. II-27]. For everyone born, departure from here is a certainty. If you have come here, you have to go one day leaving everything here. Therefore, whatever we have to achieve, we have to achieve in this life itself. Also, it makes it very clear that we did not have a human personality before we were born and we will not have this human personality when we depart from here. When you go beyond, you will have no name, no form, no body and no human personality.

What is that state? A state transcending the name and form where the human identity is dropped, it is reduced to ashes or buried. That is a state where you do not have this time consciousness of seconds, minutes and hours, days, months and years. You go beyond all that. There, there is no sunrise for you nor any sunset. In that state, you are a centre of spiritual consciousness. That is called Atman and It is a part of the cosmic, spiritual, universal Soul. That is your real identity. You are an *amsha,* a part of the supreme Being.

It is important, therefore, that you try to become aware of your real identity, try to become aware of your inner centre, your inner Being. From that awareness only, you can unfold your Divine Identity. God is Divine; and you are a part of that Divine Being. So you are also Divine. Divinity is your essential Reality. It has neither birth nor death, It has neither name nor form, It has neither time nor space, It has neither pain nor suffering. It has neither grief nor sorrow, It has neither bondage nor liberation. It is beyond all these. It is a state of peace and joy. It is a state of bliss. There is no other experience; there is no weeping and wailing, no fighting and quarrelling. All these pertain to your physical and psychological levels. All the experiences on this earth plane are either physical or psychological. But the Spirit is beyond physical and psychological levels. It is far above and physical and psychological experiences cannot reach It. It is a state which is devoid of the entire range of earth experiences of human beings. It is beyond them, It transcends them.

In that dimension, you are *Sat, Chit, Ananda*. You are the luminous Consciousness. You are the Existence and you are conscious of your Existence. That state of Being is pure Bliss, It is pure Peace and Joy. You are free and fearless.

That is to be realised. That is to be known. That is to be contemplated. That is to be experienced. That is the Goal of your life. That is the greatest ideal of real India, the spiritual India and has remained so for thousands of years, year after year. No matter what upheavals were there on the surface of India, this living stream of spirituality, the living stream *of Atma-sakhshatkara* and *aparoksha-anubhuti,* of Self-realisation and direct Experience has been flowing continuously as a silent, unseen, inner current.

You are entering the spiritual life to participate in that Experience. Apply your common sense and enquire: 'I was not a human individual before I was born. When I will have to leave this body I will be no more having this human personality. Then who am I?' Soon you will be convinced in what our seers have said: 'You are not this body, nor this mind. You are the immortal Atman. You are the shining centre of Existence-Consciousness-Bliss, you are the shining centre of peace and joy. You are full of purity, full of love, full of compassion, full of harmony, full of unity, full of light. You are *Jyotishaam-api taj-jyotih*, [That, the Light of the lights. B.G. 13/18].'

You should learn this. You must start reflecting upon your real identity. You should try to do *vichara* and *viveka;* if out of *viveka* and *vichara,* little spiritual consciousness is awakened and unfolded within you then your attending the course is really fruitful, it is worthwhile. Otherwise, you will be going back, taking with you some bookish knowledge only. No doubt you will enrich yourself and you will be a better individual. But the real purpose of Guru Maharaj in starting the Academy was not only to equip you with some knowledge of Indian philosophy, religion, scriptures, culture and all that, but also to awaken your inner spiritual consciousness out of slumber.

Yoga and Vedanta are the two terms which imply the highest Experience, the Divine Experience, the God Experience which our ancient Vedic sages, seers and *siddha mahapurushas* had experienced. That Experience is Vedanta. Vedanta is the actual Experience of your true identity; it is not merely knowledge at intellectual level. Yoga is studying and practising the scientific methods, step by step, to attain ultimately, that Divine Experience. So Yoga and Vedanta cannot be separated. They are inter-connected. The objective of Yoga is to give you the highest Vedantic Experience and in doing this, Yoga takes the totality of human personality.

Yoga tries to achieve that Experience through various paths. Through the Bhakti-Yoga, that is through the emotional path of your being, your *bhava,* sentiment, emotion, through prayer, faith, devotion and worship. Through the Raja Yoga or Dhyana Yoga with the power of your mind; by the withdrawal of the mind, making it turned inward, controlling and overcoming its restlessness and making it still, concentrating it and meditating to pierce the veil of ignorance and attaining Illumination. Through the intellectual part of your being in Jnana Marga or Vichara Marga, through *vichara, viveka,* reflection, analysis and investigation. Through the Karma Yoga by dedicating all your daily activities to God.

All these four paths can be combined into a single spiritual process, because the ultimate objective of all of them is the same. Our seers say: 'Make use of all the faculties God has given you as a part of your human personality and direct them towards the attainment of *Brahma-Jnana.*'

Yoga is a systematic way of trying to rise into the spiritual Consciousness by a well-regulated series of progressive steps. Vedanta is the ultimate Experience and Yoga is the method to attain It. So you have to learn both and make use of this knowledge and practices to bring about gradually the ascent of the spirit. It is possible provided you have the proper understanding and you are seriously practising them.

You can make use of all the four faculties by using all the four main paths—Karma, Bhakti, Dhyana and Jnana—in order to have the Vedantic Realisation. This you must bear in mind. Though you have come here as students and after going back you will be living your daily life in your own way. But even if you continue to live your life as you were living previously, the life should not be the same. Now your life is supported by knowledge and wisdom which you have acquired here. Your life may remain the same outwardly, but it will not be the same. You have a better vision, a more clear perception, you will have a fuller wisdom about your mission in life.

The quality of your life will now suddenly become transformed. It will be much superior life, much purposeful life. Your life will be ever ascending on the spiritual path. May be you will come again and again to the Ashram for the spiritual atmosphere. Life is short, time is fleeting. The human life is a very valuable, very precious gift of God. You should utilise it properly and try to attain something. Always remember what is That which is to be attained.

First and foremost, you must know how to become an ideal human individual. All the positive as well as the negative qualities are within us; many weaknesses as well as favourable qualities are within us. Whatever is not good, whatever is negative, whatever is imperfect must be eliminated gradually, so that you can shine with all that is positive and virtuous, so that you can become an ideal human individual. You will move forward continuously. Many people will be benefited by your good nature.

Secondly, our ancient ideal is to do maximum good to maximum number of people to the extent you can, in as many ways as you can, as long as you can, in all the circumstances, in all the places. Constantly make yourself a centre of *'being good and doing good.'* This body is given to you for doing good to others. The fellow human beings, animals, plants, the whole world must be benefited through your life. This is an integrated way of life. It is known as *paropakara.* We should live in such a way that we become of some use to God's creation, to all life around us.

Thirdly, your human personality is temporary. You have come here to know your real Self, to know your spiritual nature. Make use of this human birth in such a way that you can leave this body shining with Wisdom, shining with the Light of that great Experience, the Self-realisation.

These three should be your ideals:

- On the subjective level, to become an ideal human being. Make use of all your knowledge to reform yourself, so that you will have the satisfaction that you have not wasted this gift of God.

- On the outer level, to become a great *paropakari*, a benefactor of all life with utmost humility, without slightest egoism.
- On the inner level, to attain spiritual Realisation. The first two ideals are pertaining to this world; but you do not belong to this world, you belong to a higher spiritual dimension.

These three are the great tasks of human life. So, become an ideal human being. Secondly, become a centre of good to the whole world around you. While leading ideal human life, while doing maximum good to life around you as a *maha-paropakari.* You practise a rich inner spiritual life of prayer, *japa, swadhyaya,* meditation. But all these you have to do silently, secretly, in your private life. No one should know about it, do not make propaganda of it and continue to live your life as before.

This inner secret transformation should be known only to God. Suppose, you have kept in your bag some iron thing and a *siddha* turns it into gold by touching *parasmani* to it. Only you know there is an immense treasure inside but others do not know. You are careful not to tell about your treasure to anyone so as to guard it. Like that, as far as possible, never divulge your spiritual practices, your spiritual treasure to anyone. Then only, you will get maximum benefit out of it.

This is the objective of the Yoga-Vedanta Course: To help to become an ideal human individual; to transform your life with the noble ideal of *paropakara*; and with the knowledge of Yoga and Vedanta becoming a spiritual seeker, *jijnasu* and thus making your life a continuous progress towards connecting yourself with the supreme Being, a continuous spiritual ascent towards Him.

Detach, Attach

Radiant immortal Atman! Beloved children of the Divine! Immortal souls, temporarily in a state of physical embodiment and for a brief duration engaged in living with a material, physical, gross outer world, made up of ever changing, temporary perishable names and forms!

This episodic presence of yours upon this ever-changing, momentary, perishable earth plane of apparent things with names and forms is characterised by a beginning and an end. These are indisputable and incontrovertible truths regarding our presence here that it is a relationship with changeful things, perishable things, momentary in time and space, imperfect because they are not capable of imparting to you the same experience permanently. They are capable of giving you both enjoyable or painful feeling. But they cannot give you a uniform positive experience of happiness or satisfaction. You will not find it here, because of the nature of objets in this world is temporary; and temporary objects cannot give you a lasting experience.

Your life here on this earth plane is a great sojourn or a journey and you are only a traveller. You do not belong to this world. This world is not your real abode. Kabir addresses the human individual as musafir, as a traveller in this famous poem:

Utha jaga musafir bhor bhee
Aba raina kahan jo sovat hai;

[O sojourner! Awake! The dawn awaits you. There is no darkness now; why are you (still) asleep?] In another poem he says this earth plane is a halting place for a night for the caravan of life; this is not the permanent abode. People spend a night together, get themselves little aquatinted with one another, they share a common place. But when in the morning, it is time to go, each goes his way. They diverge on different paths, never to meet again.

Your journey on this earth plane is like that. But because you do not think about it, you develop attachment and get yourself involved. Attachment leads to many other situations and states of mind and the ultimate result is always pain. Attachment leads to all your problems.

When Mahatma Gandhi soon after he left South Africa and came to India and decided to devote his entire life for the service of the Indian people, he came in contact with Srimad Bhagavad-Gita and made it his personal guide-book and manual of life. He was so much impressed by the vision in this compact little scripture that he decided to make it available to all Gujarati people by translating it into Gujarati, just as, much later Sri Vinobaji translated it into simple Marathi verse which is known as *Gitaaee.* Mahatma Gandhi interestingly enough gave it the name *Anasakti Yoga,* also known as the Yoga of Detachment or the Gospel of Selfless Action.

What does Gita have to tell and teach to the traveller? Go through this world in a spirit of detachment. Do everything that has to be done, fulfil your duties discharge your obligations; but be unattached. Know that even this fulfilling of obligations and doing of duties is ultimately being engineered and carried on by some other great Power that pervades the universe and works in and through everyone of us. Arjuna had the good fortune of the actual demonstration of this by Bhagavan Sri Krishna Himself in The Yoga of The Vision of the Cosmic Form, the Eleventh Chapter. He shows to Arjuna that the whole universe is nothing but His body. There is no army, nothing; everything is finished, everything has been annihilated by the *Virat Vishvaroopa,* the all-pervading Cosmic Form. After coming down from that height of Consciousness, Arjuna says: 'Please come down to assume your normal form; I am unable to bear the impact of this Cosmic Vision.'

What we are supposed to do? The Lord says: *'nimitta-maatram bhava savyasaachin* [Be thou a mere instrument, O Arjuna! XI-33]. Be only an instrument, because I have done whatever has to be done.' It was a state of Consciousness where past, present and future are all fused into one. Arjuna sees the future events as already happened and realizes that he is only a *nimitta,* an instrument.

So, go through your life knowing fully well that some great power is working and doing all the things and you are merely an instrument of His work. Therefore, you must try to be a good instrument, an efficient instrument, so that, that work be done in a perfect manner and the Cosmic is not affected negatively by you; you have to make a positive contribution on your part. Your efficiency will suffer if you get yourself emotionally attached or get yourself involved through desire. If you are *anasakta,* totally detached, then you can act in a way that will not be affected and influenced by your personal involvement. Therefore, Mahatma Gandhi gave his Gujarati version of the Gita, the title of *Anasakti Yoga*.

This has to be the approach and attitude towards life and the manner in which to live our lives; because in the ultimate context, we have no permanent lasting connection with anyone here. A mystic says:

> *Bhai, bandhu, kutumb, parivaara, sab jeete jeeke naate hain.*
> *Kiske ho tum? Kaun tumhaaraa? Kiske bal Hari-naama bisaara?*

[Brothers, friends, family, relatives, are (all) relations of a living being only. To whom you are (truly) related? For whom you have forgotten (the only real relation,) God?] As long as you are alive, these relations are there; but once the last breath goes away, then who is your relative and who is your friend?

Therefore, go through the life unattached. Fulfil your duties and obligations in a spirit of detachment, while taking full interest in whatever has to be done. When it is being done you should see that you are not caught in the web of delusion, in the net of sorrow and suffering, and you have not to weep and wail. That is what everyone wants. That is the universal quest of all the living beings on earth, specially so, of the human being.

Because of the intelligence the man knows what he wants. But even reptiles, insects, birds, beasts, fish, who do not know and cannot understand or analyse the purpose of life and who are all after one thing, namely, avoidance of pain and suffering, have also this quest of satisfaction, contentment, happiness. So the Gita says if you do not see this clearly that you are a temporary dweller on this earth plane, that this life is a journey and you have no ultimate connection with any one here, no one belongs to you ultimately and you do not belong to anyone, if you do not ultimately cultivate this truth-oriented view of life and approach to life, you are bound to get caught in the net of delusion, in the web of *Maya* and then weep and suffer. If you want to avoid sorrow and suffering and that is what everyone wants, *be detached.*

This is the negative side of the problem as to how to conquer sorrow, how to overcome pain, misery and unhappiness. But we are also in quest of positive experience. This negative experience is obtainable even by other means also. If some one has lost everything and becomes insane, he keeps on laughing. But that is not the state we seek. This is not our goal. We should consciously, purposefully and in an intelligent manner work to avoid suffering. This avoiding of suffering takes place automatically in a certain state.

We are more interested in the other side, the constructive and the positive side of the eternal quest of all life, finding a fulfilment, attaining perfection. When there is the sense of lacking, sense of insufficiency, then you are always trying to attain fulfilment. How can this be achieved? That is the positive side of mysticism. But this has not been enough focused, enough expounded, enough explained. So the Indian philosophy has come to bear the brunt of very severe negative criticism.

There is also a negative fatalistic attitude, a world-denying view of life: No one is interested in making the world a better world. No one is interested in trying to remove the sorrows and pains of others. They say: 'This is all temporary, this will all go away. After all, everything is coated with the varnish of death. One day, everyone must leave everything and go. Therefore, why unnecessarily exert oneself?' So there is no positive effort to improve oneself. To some extent it is true. But this is due to non-assimilation of the positive dimension of our understanding of life and philosophy.

The eternal call from ancient times, as reflected in *Yoga Vasishtha* is for *Purushartha Chathushtaya,* the four-fold effort, endeavour; and it has been even built-in in our view of life. An individual human person must strive to obtain certain values to make the life worth living and to enjoy life. While you are here, why should you live in discomfort?

There was a period in human history all over the globe when monasticism, turning away from the world and turning towards God, was interpreted in a very painfully negative way. They believed that you should not give the body any comfort or any pleasurable sensation; because that experience will make you go after that, you will have a craving for it. You will become attached to that pleasurable sensation.

That concept was carried to such an unnatural, abnormal, illogical and irrational extreme, that this body was thought to be an obstacle. So they used to have certain practices of subjecting the body to torturous experiences. In summer they will wear very rough camel hair, wearing which you will die of heat. One should not have anything soft, anything cool, anything light. So inside your apparel you wore something very rough, very heavy, very uncomfortable. As though this was not enough, you slept upon a plank of wood and kept a stone as your pillow; not to use any foot-wear, to put on minimum clothes in the winter, and in this worst period even one whipped one's own self. Mortification was carried to such extremes in the medieval period.

This is the most awesome form of penance. But ultimately, nature is nature. The law of Nature cannot be flouted. The person of penance became afflicted by various diseases. He caught pneumonia, or he may get some stroke or burning of the eyes or kidney trouble and all that. But they did not attribute it to their own irrational way of extreme *tapasya*. Ultimately, their own *sadhana* suffered as the body became ill and weak. The body could not bear the strain of this extreme *tapas*. Due to the extremes, the body, the one and the only instrument through which you have to attain the supreme goal by *sadhana,* by appropriate practices, becomes no longer useful. The wrong notion of mortification beyond a certain stage becomes harmful.

This was long recognised by our ancients. They have said: *Ati sarvatra varjayet* [Give up excess of all kinds]. Equally bad is too much indulgence, too much of pampering of the body and the senses. You must know what is the limit of the other kind where if you go beyond a point your body will break down, it will suffer. You must also know the limit where if you have no longer control over the senses, the body becomes indulgent and demanding. It is through constant experimentation you can know what is your golden mean. Each one has got one's own golden mean. You must know your golden mean. There is a very telling, very meaningful saying in Tamil language which says, *Akavukku minjnaal amritham*. If you go beyond a certain measure even nectar can become poison. This is a contradiction of the very concept of nectar. Because, nectar is that which makes one immortal, deathless. Nectar can never become a cause of death. Even if a single drop of nectar gets inside you, you will never die, nothing can kill you, death can not touch you. Poison is that which causes instant death. But, it says in a funny way, if you go beyond a limit, even nectar can become poison!

In the wisdom teachings of Srimad Bhagavad Gita, this point is brought home in a very effective and telling way [VI/ 11-19]. Either way, over-indulgence destroys your chances of attainment and so also over neglecting the needs of the body. Yoga is for that person who keeps a balanced moderation in all the things. Yoga is not for that being who either sleeps too much or denies indispensable minimum requirement of sleep for the physical body. Yoga is not for that person who either eats too much or denies the body the minimum requirement of nourishment that is absolutely, indispensably required for the body.

Yoga is not for the workaholic who indulges in constant activities, over-activity, who is always distracted with some compulsive, impulsive action. He cannot remain quiet. If he remains quiet for five minutes, he will become very restless, very upset! He has to keep on doing something or other because he does not know about the inner wealth and he wants to forget the emptiness within. Nor for that being who is lethargic like a buffalo, who has no desire for activity, always lethargic; that is *tamo-guna*. Yoga is neither for that one who indulges in over-activity, nor for the one who is totally inactive, actionless.

Yoga is for that being who knows how to strike a balance between the two extremes. Therefore, we should know the golden mean, we should know how to strike a balance in all the things in our life. *Samatvam* [evenness of the mind] or moderation in all things is the key for success in Yoga, success in *sadhana*.

With regards to *Vedanta,* they say: 'Within you, you must always have that *bhava,* that attitude that nothing exists; this world is only a dream appearance, there is no reality in it.' You must always cherish this *bhava.* Then only, you will be safe and you can be successful in overcoming all the external impacts and influences. This attitude should always be there within you, it should never be lost. But you should not apply it to your outer relative plane of worldly relation; otherwise it will cause problems. Also, even if you apply this *Mayavada,* this *mithyavada*—everything is *mithya*—to all the three worlds, you should never apply it to your Guru! Don't say he is also *maya,* he is also *mithya.* You must serve him, surrender to him and be like a slave, like a servant to him.

Advaitam trishu lokeshu an-advaitam gurunashah. [Non-duality with all the three worlds, but not with Guru]. You can apply this principle, 'I alone exist, nothing else exists besides me,' towards all the three worlds but not to your Guru.

I used to explain it in this way: The *Advaitic* declaration is based upon the non-dual, transcendental, absolute Experience that goes beyond relativity. That was the Experience that was revealed. It was not meant for application in day-to-day life. To explain this, I used to say: Some modern, superb, great, eminent physicists have discovered there is no such thing like real, solid matter. This is all energy, waves of energy. They call it energy in waves. That means, they have come to the view that the real, gross, visible, tangible, solid matter is only our imagination due to the sense perception; there is no such reality. There is no gross, solid matter.

They first thought that the ultimate substance is an atom. But they went beyond and found that even this atom which was once irreducible, smallest of all particles, is divisible, and there are smaller parts within the atom: proton, neutron and electron. When they looked into this electron with modern super-instruments, they saw that even within this electron there is ninety per cent empty space only! There particles are moving at tremendous velocity. So they give the appearance of some tangibility or solidity, but otherwise, it is more of empty space than actual substance or matter.

In this way, ultimately, they came to the conclusion of the Shaakta Philosophy that everything is nothing but Shakti. We have the concepts of Adi Shakti, Kriya Shakti, Para Shakti, the Primal Cosmic Power, the motivating Power, and the Transcendental Cosmic Power. The modern Physicists have come to similar discovery and they are trying to make it intelligible to the layman. This is the actuality about this world. That is what they have actually seen, experienced, and ultimately arrived at as the conclusion of modern super-physics.

In the same way as in metaphysics, is that tremendous Experience, the towering pinnacle point of ultimate transcendental, absolute Experience of *Ekameva-advitiyam Brahma* [Brahma is one alone without a second. *Sadachara-anusandhanam,* 44] and *Brahma satyam jagan-mithya.* [Brahma alone is real, the world is unreal. Brahma-jnanavali-mala, 20] That is something which you must keep as your Goal: 'I have to go and attain That, I have to reach that state of non-dual, cosmic Consciousness.' Then there will be no more duality. Pleasure and pain, happiness and misery, joy and sorrow, and all such duals become meaningless, they are at a level that is far below; whereas you go beyond all the relative or opposite dual experiences. You should say: 'Yes, that is the Goal of my life. That is the only Goal to be attained. I must reach that state of non-dual Consciousness. But it is not a principle to be applied to my day-to-day living. Otherwise, unnecessary chaos will arise.'

I used to try to bring home this unavoidable and indispensable axiom of our behaviour in this world by an analogy: Modern physics has come to this discovery, this conclusion, that there is only energy, there is no real, solid, substance called physical matter. Suppose you are in a room and you want to go out. There are a couple of doors. If you say, 'Why should I use the doors? They are nothing but energy. There is no real thing called matter. So let me walk through the wall.' If you try to walk through it, it will hit on your head. Though their discovery, their ultimate declaration, is an undisputed fact, yet it is not a fact which you can bring down to your field of daily relative activity. Then you will have to go and get yourself bandaged. Nevertheless, your breaking your head in going through a wall does not contradict or invalidate the truth of the discovery of modern super-physics. It is true but it is not for application in your daily life.

In the same way, *Advaita bhavana* is true in one dimension; but it is not for application. Even that great philosopher Jagadguru Adi Sankaracharya, had to posit and include within his philosophical system, a secondary reality called relative reality. *Brahma satyam, Jagan-mithya. Brahma* is the *kevala Satta,* the absolute Reality. True. But as long as you are bound up in the worldly consciousness, it has also a relative reality, a reality which you can not dismiss, which you cannot ignore. It is known as *Vyavarharika Satta.*

While you are in a state of dream, dream alone is real; outer world does not exist. It is only when you wake up that the world becomes real and the dream world becomes unreal. Similarly, as long as you are here, in your present state of consciousness, this world which has a relative reality, is true and is relevant to you. With reference to your own present state of relative consciousness, where your experience is not non-duality but on the contrary, where your experience is the multifariousness, you do have to make a distinction between red and blue, green and yellow, white and violet. When you are driving a car and on an intersection the red light comes, you cannot drive through it saying that you believe in non-duality and for you there is no distinction between red and green. You will get arrested. So we have to accept the relative reality of the world.

When you have to climb up a staircase you have to use your limbs in one way but when you have to climb down the same staircase you have to use your limbs in a different way. Likewise, you have to use different yardsticks for the two different states. You cannot say: 'I will drink my solids.' It will get stuck and you will get choked. You cannot escape giving the due attention and regard to the relative reality that is very much a fact. As long as you yourself are in a relative state of consciousness, you feel: 'I am a male, I am a female, I am young, I am old,

I am tall or I am short.' If you are tall, you can reach up and take something from a height; but if you are short, you cannot say: 'All these notions are false, I am the Atman.' You have to ask for a stool, get up on it and then take the thing.

This being the case, this world has to be dealt with on its own level, on its own dimension of reality. Having come into this world, we should know that all the things are limited, changeful, perishable. Therefore, no lasting experience of true happiness can be found here. This is only fifty per cent of the truth. Not only happiness is not to be found here, on the contrary, what you get here if you get involved, is only weeping and wailing, grief and sorrow, disappointment and frustration, agony and misery, little fulfilment, then again frustration and disappointment!

What is the way to avoid it? Be *detached.* Don't get caught in the net of attachment.

What is the way of positive experience? This is where the constructive and creative side of Yoga-Vedanta comes. Does it simply say: 'Don't go in that direction. Don't get involved. You will get caught in sorrow, you will weep and wail.' Does it leave you in the lurch like that? No, no, no! It also says; 'Go in that direction, you will get total bliss, total satisfaction, eternal happiness. You will get all that you want and even more than what you ever imagined you will get!'

How to do it? What are the means? Well, that is the real question.

Here I would like to give you two oft-quoted analogies. One is an analogy from Vedanta, and the other is an analogy which you can imagine from your common sense.

Someone was imprisoned on a tower because he had displeased the ruler of the land. His hands and feet were bound by a rope and he was thrust up like a chicken, thrown up on the tower. While coming down, the jailer closed the door of the tower. No one had any access to the tower. He was left to die exposed to the wind, heat, cold of the night and may be vultures later on. He was in such helpless condition; but he was not a person to despair. As an optimist, he thought: 'There is nothing hopeless. Even in the most extreme conditions, there may be some way out. There is always some hope.' So, as he lay there tied in that helpless condition, it struck him: 'So long as I go on struggling with this bond, the tightness will not relax. I must relax and make myself as small as possible. Then only some part, some little bond may become a little loose.' Several hours passed.

Whole day, he had not taken anything. His belly was now empty. That also became a blessing in disguise. He thought: 'Now is the time. I will try to relax.' He relaxed his muscles, he tried to make himself as small as possible. He drew his belly in, and slowly tried to work upon the bonds. At long last, he managed after a very, very careful gradual effort to bring out one bond. From that point onwards, he went on loosening his body. After much effort, exhaustion, perspiration and straining, he managed to gradually remove all the knots that had bound him. The rope which had made him a helpless prisoner, lay at his feet to rescue him.

When it was totally dark, he tied one end of the rope to some portion of the tower and then let him slowly slip down the entire length of the rope. At the end of the rope he hung himself and gave up even that rope which had rescued him, and jumped on the ground to become liberated for ever.

Suppose, a man is walking in a forest and a sharp thorn enters his foot. It is so painful that he cannot walk further. He is not equipped with a thorn-puller, a needle or any sharp instrument. There were many thorns around. He picks up a strong, sharp thorn. He uses it as an instrument to pull out the thorn from his foot. The cause of his pain and agony was a thorn; and a similar thorn comes to his rescue! Now, he discards both the thorns and moves ahead on his path.

What causes the bondage of the *jivatma?* Our ancients say the cause of the bondage of *jivatma* is attachment, desires, craving. Once you are caught in this net of *Maya,* you are bound. Now your desires go on increasing; there is more and more craving for this thing and that thing, for money, wealth, position, power, authority, status, enjoyments, name and fame, appreciation and all that. You run after so many things! Anything you run after, catches hold of you, you become possessed by it, you become enslaved by it. Our scriptures tell us that if you are caught by even one of the five senses, you are doomed.

A fish is caught in the hook due to its greed of taste; a moth burns itself into the flame due to its attraction of *roopa,* the outer appearance; a deer is killed because of its attraction for sound of music, an elephant falls in captivity being blinded by touch; and the attraction of smell destroys the bee. Various creatures get caught by one or the other senses. But in the poor human individual all the five senses are centred! He is enslaved by all the five senses! When a single sense is enough to lead to your destruction what will be your fate when all the five attack you? So, be very careful and guard yourself from being enslaved by the senses.

You are caught in this net of *Maya,* in this worldly process because your mind conceives desire, because your mind gets attached. It is the mind which is doing *avichara* and *aviveka,* improper reasoning and lack of right discrimination. It is the mind which says, 'I want this thing and I want that thing; I want to enjoy the things; my happiness is in these objects of attractions. I want status, I want position, popularity.' The mind misguides you and pollutes your intellect also. Your intellect is clouded due to lack of proper enquiry and discrimination. That is the second factor that contributes to your bondage.

Then comes your involvement in the ceaseless activity, driven by the desires and attachment, emotions and sentiments. This is another dimension of the mind.

So long you lack *viveka* and *vichara,* your mind moves in the wrong direction. Your personality has four aspects: The physical level, the emotional level, the mental level and the intellectual level. In the absence of *viveka* and *vichara,* you are caught in the net of the outer world process of the phenomenal existence through all these four aspects of your nature. This is the situation. But the thorn which was the cause of the problem, becomes the solution of the problem; and the same rope which bound the man, becomes the means of his freedom. So, applying the solution of these analogies, you should make use of all your four faculties to attain freedom by diverting them Godwards. You are the highest of all the creations of God and you have been endowed with the power to think and feel, reason, and act in a purposeful way. But, unfortunately all of them are applied in a negative manner in a negative direction.

You have to reverse that process. Your attachment to persons and to the passing, perishable objects of pleasures and enjoyments is the cause of your delusion and bondage. Now make use of the same faculties and attach yourself through intense longing, through intense devotion, intense emotion, to the eternal Reality. Put your reasoning onto the right channel through enquiry and discrimination.

Analyse the real nature of this world and know that there is not an iota of real happiness in these objects of attraction which tempt you in sense indulgence. They are only source of pain and sorrow. Yet they always tempt you, draw you towards them, make you run after them, enmesh you, entangle you, enslave you. Know well that they are momentary and useless, there is no real worth in them. They are *alpa*. How can you have any attachment once you know the real nature of the world?

So making use of the intellect in the right way, in a rational, logical way, with proper discrimination, you can convert the very bondage into an instrument of freeing yourself from the bondage of the outer world through the same mind with the process of *vichara* and *viveka*: 'What is the Reality beyond? What is it that can make me really satisfied, happy, free, full of peace? That is only in the non-dual Experience. *Dvaitad vai bhayam bhavati.* Where there is duality, there is fear. If there is no duality, there is no fear. Who is to fear whom, if there is only One non-dual Consciousness pervading all?'

Vedanta says, through intellect, through proper reasoning, proper discrimination, proper analytical thinking, come to the conclusion that there is only one Reality, the non-dual Consciousness of the Supreme Reality alone. There is Absolute

Peace, Absolute fearlessness, Absolute joy, eternal happiness in That. In that way, attach yourself to That. And through your *bhava,* through your sentiment, either through feeling or through reasoning or through giving a spiritual turn to all activities, always feel: 'I am always in the presence of God. Therefore, all activities are adoration to that constant Presence, in whose presence I live, move and act.'

If the power of thought is completely diverted towards and focused upon ephemeral things, ultimately, it will become a costly bondage. If the same power of thought is directed towards the Great Reality, being taken away from the passing scene, turned round and given a new direction; and the outgoing power of the mind is made to reverse its direction, made indrawn, through *pratyahara, antarmukhatva,* [withdrawal of the mind from the sense objects and focussing the mind inward] then the scattered rays of the mind are focused within through discipline, through the subsiding of this constant restlessness of the mind by ceaseless practice.

Raja Yoga is the *yoga* of the mind, *yoga* at the mind level, *yoga* of the thought-power. Its *sadhana* is in completely subsiding this constant restlessness of the mind and then gathering the rays of the mind inwards into one powerful ray and centring it upon the concept of the Great Reality, making it steady and focusing it upon It like 'a lamp placed in a windless spot does not flicker (B.G. VI-19),' by constant practice again and again. Whenever the mind moves away, bring it back and once again focus it upon the Great Reality.

Then one day you will succeed. The mind will stop moving out. It will lose its restless nature. It will change its nature and will become your ally, become your asset. It will no longer be your liability or your problem. It will be your solution. It will become your asset. A concentrated mind focused upon the supreme Reality will ultimately bring you to a state where the outer world fades away, recedes and becomes centred in the Reality through the concentrated power of your mind.

So that is Yoga of the mind. Vedanta is the Yoga of intelligence, rational intellect, reasoning process, Vichara

Marga. Prema Marga or Bhakti Yoga is *yoga* of the heart, *yoga* of the emotions, yoga of all your feelings, yoga of sentiments. Karma Yoga is the Yoga of connecting all your activities from morning till night, with God, not with a remote Reality but here and now Reality, because God is all-pervading, God is present here all the time.

The world is always filled with God's presence. Connect all your activities with God manifested around you, then it becomes *yoga*. Action will not take you away from God, it will take you towards God. Offer all your actions to Him, and your whole life will become spiritual. All the outer activities of the day-to-day life will also now take you in that very same direction in which your devotion is trying to take you, in which your concentration is trying to take you, in which your Vedantic Vichara Marga is trying to take you! Then what happens? Your entire life becomes integrated and takes the form of the Absolute. This is the rationale, this is the scientific basis of the four Yogas, namely, Karma Yoga, Bhakti Yoga, Dhyana Yoga and Jnana Yoga.

Thus the very situation which is now your problem, is now holding you bound, that very situation becomes converted into a liberating process when you divert and channelise all these potentials of your human nature in that direction, Godward direction. Now your whole personality and all your actions become God-oriented. You are having in yourself, all the ingredients for your liberation. You don't have to search outside. God does not expect you to do something, with something which do not belong to you, which you don't have, which He has not endowed you with. Whatever has now become your problem, the cause of your bondage, that very thing becomes the means of your liberation, by taking away from here and diverting it in that direction.

At the end of his life Gurudev gave two words to sum up the entire Yoga and Vedanta: DETACH, ATTACH. *Maya Ka sang Todo, Hari ke sang jodo.* [Detach yourself from *Maya*, attach yourself to the Lord.] That is the Yoga-Vedanta *sutra* for you.

Ponder deeply over this and try to find out how you can do this process of gradually detaching from here and attaching it There, giving the Godward direction and keeping up in that direction. In that lies your sure Liberation. Let all your faculties become oriented towards It. Let

Asato maa sad-gamaya;

Tamaso maa jyotir-gamaya;

Mrityor-ma Amritam gamaya

God bless you all. Hari Om Tat Sat.

4

The Institution of Shankaracharya

Shankaracharya, (IAST: *Zankarchrya*) is a commonly used title of heads of machas (monasteries) in the Advaita tradition. The title derives from Zankara of Kaladi, a theologian of Hinduism, who established four machas in the four regions of India. As the first historically recognized teacher in the lineage, Zankara is also called di Zankara (the first Zankara). The four institutions traditionally said to have been established by di Zankara are:

- The *Uttarmnya Matha*, or northern Matha at Joshimath.
- The *Pkrvmnya Matha* or eastern Matha, the Govardhana Matha, at Puri.
- The *Dakshinmnya Matha*, or the Sringeri Sharada Peetham, the southern Matha, at Shringeri.
- The *Pachimmnya Matha*, or the Dwaraka Pitha, the western Matha, at Dwarka.

The heads of these four institutions are considered the principal Shankaracharyas in India today. Kanchi Matha claims that Adi Shankara established it as the *Sarvagnya Peetha* in a position of supremacy over the other Mathas of the subcontinent, before his samadi. However this claim is disputed by the other four Peethams.

In addition, the heads of the Sumeru Matha in Varanasi, the Karavir Matha in Kolhapur, Maharashtra, the Shankarananda Matha in Puri, Orissa and other institutions are also referred to as Shankaracharyas. There are more than 100 such Shankaracharyas in India today.

Vedanta: Highest spritual knowledge knowing which, there remains nothing further to be known. It is Self knowledge (*Atmavidya*). It is knowledge of the Absolute Truth (*Brahmavidya*). Vedanta teaches the real or essential nature of God, the universe, and the individual being (Self) and its oneness with God. The quintessence of the teachings of Vedanta is that Brahman (God) which is Existence, Consciousness and Bliss Absolute is the only Reality and that the universe is an illusory appearance like a mirage seen in a desert or a rope mistakenly seen as a snake, and that the individual being (self) is essentially Brahman or God Himself.

This oneness of the self (*Jiva*) with Absolute Consciousness (Brahman or God) is the goal of Vedanta. For want of the right understanding about his real nature, man remains deluded and considers himself limited, unhappy, miserable and beset with innumerable problems. Vedanta shows the way to eradicate these problems once and for all and enables man to attain the highest spiritual enlightenment, peace, happiness and freedom (*moksha*) from repeated births and deaths. Knowledge of the Self (*Atmavidya*) and knowledge of the Absolute Consciousness (*Brahmavidya*) is the highest knowledge to be obtained. It is the highest of all human pursuits.

The Jyotirmath Shankaracharya Lineage in the 20th Century

There is currently a succession dispute at the Jyotirmath Shankaracharya seat, the origin of which dates back to the year 1953. Till recently, the two major rivals were Swami Swarupananda Saraswati (who is also the Shankaracharya of Dwaraka in the west) and Swami Vasudevananda Saraswati. Since 1993-4, another Sannyasin named Madhava Ashrama has been a third claimant to the Jyotirmath title.

Background: Adi Shankaracharya (8th century) is traditionally said to have established four Mathas (monasteries) in India, and to have placed them under the leadership of his four chief disciples. The heads of these four and other monasteries of the Dasanami orders have come to be known as Shankaracharyas themselves, in honour of the founder. They are considered to be the leaders of the ten orders of the *Dasanami Sannyasins* associated with Advaita Vedanta.

The principal eastern (*Purvamnaya*), southern (*Dakshinamnaya*) and western (*Pascimamnaya*) institutions are based at Puri (Orissa), Sringeri (Karnataka) and Dwaraka (Gujarat) respectively. The northern (*Uttaramnaya*) Shankaracharya seat is at Jyotirmath (also known as Joshimath) near Badrinath. In addition to these four, there are numerous other Mathas throughout India, and the seven Dasanami *Akhadas* (Juna, Niranjani, Mahanirvani, Ananda, Atal, Avahan, Agni—the last is a Brahmacharin *Akhada*, not a Sannyasin one) that have their own separate administrations and leaders.

The history of Jyotirmath is extremely complicated. According to official accounts, after the period of one Swami Ramakrishna *Tirtha* in the 18th century, the Matha was extinct for about 165 years, before it was revived in 1941, under Swami Brahmananda Saraswati. However, in the meantime, various Sannyasins had claimed to be the Shankaracharya of Jyotirmath, and for some time, many people thought that the *Rawal* (head-priest) of the Badrinath temple was also the Shankaracharya of Jyotirmath.

There are records of lawsuits from the early 1900's, which show a number of names, each laying claim to the Jyotirmath Shankaracharya title. Before discussing the recent history of this institution, one issue must be properly understood. What does it mean for a Matha to have become extinct? In the traditional understanding, the answer is always with reference to the Mathadhipati, the head of the Matha. To say that a Matha has become extinct is to say that the lineage of successors has petered out, either because of confusions in the line of succession or because of lack of qualified people to lead the Matha.

The appointment of Swami Brahmananda Saraswati in 1941 was made by a group of monks and pundits based in Varanasi (the *Bharata Dharma Mahamandala*, *Kashi Vidvat Parishad*, and the *Akhila Bharatiya Dharmasangha*), with the blessings of Swami Bharati Krishna *Tirtha*, the then Shankaracharya of Puri. Swami Chandrasekhara Bharati of Sringeri also endorsed Brahmananda's appointment. Thus, right from the beginning of the Jyotirmath's revival in 1941, the opinions of the heads of other Amnaya Mathas were taken seriously into consideration.

Another important factor that legitimated the Jyotirmath revival should not be overlooked. This is the involvement of the Hindu kings in north India in the process, and their acceptance of Brahmananda Saraswati. The ruler of Garhwal was important because Badrinath was in his province, while the rulers of Varanasi and Darbhanga were well known and respected as patrons of several religious institutions. India was not yet independent in 1941, and this royal recognition helped in overcoming potential opposition from previous claimants to the Jyotirmath title, and their followers.

Brahmananda was selected as he was widely regarded as the epitome of the qualifications mentioned in texts like *Mahanusasana* and *Mathamnaya*, which are attributed to Adi Shankaracharya. For the purposes of this discussion, it is immaterial to investigate who wrote these texts, or to question whether Adi Shankaracharya established any Mathas at all. It is sufficient to note that the living tradition of Advaita monasticism more or less unanimously accepts these texts and the four Amnaya Mathas as originating from Adi Shankaracharya himself.

Beginnings of Conflict: Swami Brahmananda Saraswati passed away in 1953, but he had not clearly indicated his successor. This immediately caused a problem, as he had initiated a number of disciples into Sannyasa. A few weeks after he passed away, a will was found, according to the terms of which, a disciple called Swami Santananda Saraswati was named as the first choice for succeeding to the Jyotirmath title.

However, many followers of Brahmananda Saraswati were satisfied neither with the credentials of Santananda, nor with the validity/authenticity of the will. Perhaps, the doubts about the will were themselves based partly upon the perception that Santananda was not a good choice for successor. His quickness to take charge of the Matha administration on the basis of this will also probably raised many eyebrows. Meanwhile, there was a widespread rumour that Brahmananda Saraswati had been poisoned. This set a number of civil lawsuits into motion. To the best of my knowledge, no criminal lawsuits were filed against anyone, on the basis of the poisoning theory. However, Santananda's reputation definitely took a blow, although the major complaint against him was simply that he was unfit for the post of Shankaracharya, because he did not measure up to the qualifications described in the *Mahanusasana* texts.

Also, in 1953, Swami Hariharananda Saraswati (popularly known as Karapatri Swami), another disciple of Brahmananda Saraswati, was seen as the more deserving candidate to become the Shankaracharya, but he didn't want the title. In fact, as the head of the *Akhila Bharatiya Dharmasangha*, it is said that Hariharananda had been the first choice for the Shankaracharya post in 1941, but he had declined and proposed his guru's name (Brahmananda) instead. It is also said that it was Hariharananda who convinced his guru to accept the position. Hariharananda Saraswati passed away recently, and avoided the Shankaracharya title for himself, but the opinion of his followers is reflected in the title they have given him-*Abhinava Shankara*.

Divided Lineage: Because of the controversy over Brahmananda's will and Santananda's succession, the organizations involved in reviving Jyotirmath in 1941 considered other nominations for the Shankaracharya post. These efforts were blessed by Swami Abhinava Sacchidananda *Tirtha*, the then Shankaracharya of Dwaraka. In 1953 itself, one Swami Krishnabodha Ashrama was appointed as the new Jyotirmath Shankaracharya, contesting Santananda's claim.

Krishnabodha Ashrama was not a direct disciple of Brahmananda Saraswati, but given the nature of Jyotirmath's revival in 1941, this was not necessarily a disqualification. The new appointment also had the support of the Puri Matha, but it must be noted that this Matha was to have a few succession problems of its own, within a decade. When Krishnabodha Ashrama passed away in 1973, he nominated Swami

Swarupananda Saraswati to the title. Swarupananda is a direct disciple of Brahmananda Saraswati, but he has also studied under both Krishnabodha Ashrama and Abhinava Sacchidananda *Tirtha* of Dwaraka. Meanwhile, Santananda had not relinquished the Shankaracharya seat, so that by this time, the two major claimants of the title were Swarupananda and Santananda.

In the year 1980, Santananda stepped down from the title, in favor of Swami VishnudeVananda Saraswati, another disciple of Brahmananda. However, VishnudeVananda Saraswati passed away in 1989/90, while Santananda Saraswati was still alive. Following this, one Swami Vasudevananda Saraswati was named as the successor. Santananda passed away in late 1997, and Vasudevananda Saraswati is currently the sole representative of this lineage.

Vasudevananda was present at the appointment of a Mahamandaleswara of the Mahanirvani *Akhada* in 1995 (according to Hinduism Today, August 1995). Adding to the complexity of this dispute is the fact that according to the terms of Brahmananda's contested will, one Swami Dwarakesananda Saraswati was supposed to have been the second choice after Santananda. There is no indication that Dwarakesananda ever claimed the Shankaracharya title, or that it was ever formally offered to him. A similar situation obtains with a Swami Paramatmananda Saraswati, who was also named in the will, but as the next choice after VishnudeVananda.

It is difficult to ascertain the opinions of the heads of other institutions about this course of events. After the nomination of Krishnabodha Ashrama, the Puri and Dwaraka Mathas do not seem to have been involved in Jyotirmath affairs to any significant extent. In the south, Swami Chandrasekhara Bharati of Sringeri had been succeeded by his disciple, Swami Abhinava Vidya *Tirtha*, in late 1954.

Some official Sringeri accounts mention that the new Swami met Santananda Saraswati during his first northern tour in 1956/7, but this may not indicate that the Sringeri authorities differed from the Puri and Dwaraka authorities, with respect to endorsing Santananda. It remains unknown whether the Sringeri Shankaracharyas tried to mediate or took sides in the Jyotirmath dispute in this early period. However, in 1979, when a conference of the Shankaracharyas of the four Amnaya Mathas was held at Sringeri, Santananda and VishnudeVananda were not invited. It was Swarupananda Saraswati who represented Jyotirmath. There is no indication that the rival lineage of Santananda and his disciples was endorsed at this time by any of the other Shankaracharyas.

Accounts written by Swarupananda's followers do not mention Krishnabodha Ashrama very prominently, probably because he was not a direct disciple of Brahmananda Saraswati. Swarupananda, being a direct disciple of Brahmananda, traces his lineage directly to the ascetic who was the first Shankaracharya of the revived Jyotirmath. Needless to say, the list of Shankaracharyas of a Matha must be distinguished from the *Guru-Sishya* lineages of the Sannyasins who become Shankaracharyas. In an ideal situation, the lineages are identical, but circumstances often dictate otherwise.

Other major factors that affect this succession dispute are the relationships of the principals with Indian political parties and with internationally popular gurus. Swarupananda Saraswati's involvement with the Indian Congress party dates back to the period of the Indepedence struggle, before he became a Sannyasin. He remains close to numerous Congress politicians (*e.g.* Digvijay Singh of Madhya Pradesh and P. V. Narasimha Rao, the former Prime Minister of India), and has been quite vocally anti-VHP and anti-RSS.

In contrast, or perhaps because of this, Santananda, VishnudeVananda and Vasudevananda have all had the support of the "Hindutva" organizations. Vasudevananda is usually present at major VHP and RSS events, where he is introduced as the Jyotirmath Shankaracharya. Santananda and VishnudeVananda have also had close connections with Mahesh Yogi, who used to be Brahmananda Saraswati's secretary. In fact, the earliest doubts about the will left by Brahmananda Saraswati were linked to suspicion of the motives and actions of Mahesh Yogi (then called Mahesh Brahmachari). These TM connections probably did not endear Santananda and VishnudeVananda to the predominantly Brahmana following of the various Mathas. I have also heard rumours that when Santananda stepped down in favor of VishnudeVananda, there was the hand of Mahesh Yogi behind it. As with so many aspects of this dispute, I don't know if this is just rumour, or if there is something more to it.

It is well known that Swarupananda and Mahesh Yogi don't see eye to eye on any issue, and the *ex*-TM literature has much information about their disputes. However, the connection of Mahesh Yogi to Santananda and his lineal successors is not without its own complications. For example, Deepak Chopra, the popular New Age author, who used to have intimate ties to Mahesh Yogi and his organization, has now broken his connections to him, and claims acknowledgement directly from Vasudevananda Saraswati instead.

Other Candidates: Another Swami has claimed that he was once offered the Jyotirmath Shankaracharya title, which he respectfully declined. This is Prakasananda Saraswati, who has set up an "International Society of Divine Love" and a Rasesvari Radharani temple, known as Barsana Dham, in Texas, USA. Although he was initiated into Sannyasa by Brahmananda Saraswati, his personal religious philosophy is Achintya Bhedabheda, associated with Chaitanya Mahaprabhu and Gaufdiya Vaishnavas.

This leads me to seriously suspect his claim of having been offered the Shankaracharya post. Still, for the record, if his claim is valid, then we have two Swamis who have rejected offers to become the Shankaracharya of Jyotirmath, namely Hariharananda Saraswati and Prakasananda Saraswati. We should also not forget Dwarakesananda Saraswati, and Paramatmananda Saraswati, who were named in a contested will, but did not really get an opportunity to press their claims.

And if we dig deep enough, we will probably unearth more disciples of Brahmananda Saraswati, who were all potential candidates for the Shankaracharya position at one time or the other, which they all rejected. As an aside, a few relatively unknown, early Indian publications of the Transcendental Meditation movement claim the Shankaracharya title for Mahesh Yogi himself, but this is not to be taken seriously. While Mahesh Yogi was indeed one of the followers of Brahmananda Saraswati, he has never formally become a Sannyasin. Moreover, his international following probably obviates any need for him to claim the Shankaracharya title for himself now.

Swarupananda at Dwaraka: A further complication was introduced in 1982, when Abhinava Sacchidananda *Tirtha*, the Shankaracharya of Dwaraka, passed away, leaving a will with a few names as possible choices for his successor. Among these was Swarupananda Saraswati of Jyotirmath. The others either declined or were eliminated from consideration for one reason or the other. Swarupananda was then coronated at Dwaraka, in a ceremony presided over by Swami Abhinava Vidya *Tirtha* of Sringeri. Swarupananda Saraswati's Dwaraka title is undisputed, and he is routinely described in press reports as the Shankaracharya of Dwaraka.

However, it should be noted that while his status as the head of two principal Mathas is somewhat unusual, and also confusing for the lay public, it had not been insisted that he relinquish his position at Jyotirmath, before taking charge at Dwaraka. Swarupananda had attended the 1979 meeting of the Shankaracharyas at Sringeri, in his capacity as the Shankaracharya of Jyotirmath. He also attended the funeral ceremonies of Swami Abhinava Vidya *Tirtha* at Sringeri in 1989.

In June 1993, a joint statement was issued by the Shankaracharyas, in connection with the Babri Masjid demolition, which Swarupananda Saraswati signed twice, in his dual capacity as the head of both Dwaraka and Jyotirmath. And as Jayendra Saraswati of the Kanchi Matha has also signed the 1993 statement, I assume that he too accepts Swarupananda at both Dwaraka and Jyotirmath. Swarupananda has publicly stated that the Kanchi Matha is only a branch of the Sringeri Matha, but he did attend the birth centenary celebrations of Chandrasekharendra Saraswati of

Kanchipuram in 1993. It must be noted that the Kanchi Matha is a very influential institution today, and although it is not one of the four original institutions, the opinion of its head counts for something in these controversies.

Indeed, the very presence of Jayendra Saraswati along with the heads of the four Amnaya Mathas is a marked change from the absence of the Kanchi Matha in the 1979 meeting of the Shankaracharyas, and is an acknowledgement of the current political importance of this institution. Clearly, at least in the eyes of these others, Swarupananda Saraswati is the Shankaracharya of Jyotirmath, and also the Shankaracharya of Dwaraka. However, Vasudevananda Saraswati's connections with the *Akhada Parishad* (a coordination body that deals with such matters as order of procession of the *Akhadas* during the Kumbha Mela, etc.) indicate that the Dasanami *Akhada* structure may not be unanimously supportive of Swarupananda.

Madhava Ashrama's Claim to the Title: There is a third claimant to account for, named Swami Madhava Ashrama, who is also a leading light of the *Akhila Bharatiya Dharmasangha*. This Sannyasin is a disciple of Krishnabodha Ashrama, whose name has been noted earlier. Madhava Ashrama's name crops up routinely in Nepali newspapers like The Kathmandu Post and The Rising Nepal. He visited Nepal in 1997, for the 25th anniversary celebrations of King Birendra's accession to the throne. He also attended the Kumbha Mela in 1998, where he was injured in a riot.

Madhava Ashrama's claim to the Jyotirmath Shankaracharya title is based on the stance that after Swarupananda Saraswati took up the Dwaraka Shankaracharya title in 1982, his title at Jyotirmath has been nullified by the passage of time. Supporters of Vasudevananda's claim have also offered an identical argument. Madhava Ashrama does not recognize the claim of Vasudevananda Saraswati to the Jyotirmath title, as he traces his own claim to the title through his guru, Krishnabodha Ashrama, and to the decision of the *Kashi Vidvat Parishad* and the *Akhila Bharatiya Dharmasangha*.

According to his followers, in 1982 itself, Niranjanadeva *Tirtha* of Puri had requested Swarupananda to give up his Jyotirmath title in favor of another Sannyasin before taking up the Dwaraka Shankaracharya post. However, for more than a decade after assuming charge at Dwaraka, Swarupananda Saraswati had continued to retain his Jyotirmath title. It is said that in 1993-4, the *Kashi Vidvat Parishad* and the *Akhila Bharatiya Dharmasangha* decided that this was creating much public confusion, and decided to appoint Madhava Ashrama as the Shankaracharya of Jyotirmath. The coronation ceremony is reported to have been conducted in Varanasi itself. To summarize, the three competing Jyotirmath lineages are:

1. Brahmananda Saraswati (1941-1953; revived the Matha after a vacancy of 165 years)		
2. Santananda Saraswati (1953-1980, d. 1997)	2. Krishnabodha Ashrama (1953-1973; not a direct disciple)	
3. VishnudeVananda Saraswati (1980-1990; disciple of Brahmananda; predeceased Santananda)	3. Swarupananda Saraswati (since 1973; disciple of Brahmananda) (since 1997 [1990];	3. Swarupananda Saraswati (1973-1993/4 [1982])
4. Vasudevananda Saraswati disciple of Santananda)		4. Madhava Ashrama (since 1993/4; disciple of Krishnabodha Ashrama)
	Not including-Hariharananda (Karapatri), Dwarakesananda, Paramatmananda and Prakasananda.	

The Role of Other Traditional and Modern Institutions

The above discussion has dealt only with the cultural and religious aspects of the Jyotirmath dispute. A few centuries ago, such problems would have been referred to the local king, and perhaps solved quickly. In independent India, the dispute has been taken to the secular courts, but these are quite different from the old princely *durbars* in their procedures and rules. The judges also lack legislative and executive authority over religious institutions, unlike the Hindu king of old days.

It seems to me that those who did not wish to acknowledge Santananda as the Shankaracharya of Jyotirmath did not sufficiently appreciate these changes in modern times, and expected the courts to accept their cultural, moral and religious arguments as legally valid. Thus, none of the civil suits in this dispute seems to have been framed in terms of contesting the legal bona fides of Brahmananda's will. Consequently, although one judge did acknowledge the merits of the claim that Santananda was not properly qualified, he found no legally valid reason to give a verdict voiding Santananda's claim to the title.

In 1980, after Santananda abdicated in favor of VishnudeVananda, a new lawsuit was filed, on the grounds that according to Brahmananda's will, Dwarakesananda Saraswati should have been appointed in case Santananda stepped down. Thus, notwithstanding what was privately thought about the will and its legitimacy, its terms were co-opted, as a strategy to displace VishnudeVananda. However, the proverbial judicial delays in India played their role, and before this case was even heard, Dwarakesananda passed away. After VishnudeVananda also passed away, and Vasudevananda took charge, a fresh round of litigation was begun. The latest development from this angle is that on February 22, 1999, the court at Allahabad, Uttar Pradesh, has passed an interim order, prohibiting Vasudevananda from using the title of Shankaracharya to collect any donations, till the legal case gets resolved either way.

The reasons cited for this order are that his installation is suspect, and that Swarupananda has the better claim to the Jyotirmath title, both for historical reasons and by virtue of being acknowledged by the other Shankaracharyas. Thus, notwithstanding the previous legal standing of Santananda at Jyotirmath, his disciple and successor seems to have suffered a legal setback at present.

According to currently applicable Indian law, which is itself based on pan-Indian tradition, a Mathadhipati has a right to choose his successor from among his disciples. In an ideal case, this right is vested solely in and exercised solely by the Mathadhipati, in accordance with the norms specified in texts like the *Mahanusasana*. However, given the history of this particular case, the involvement of two bodies external to

Jyotirmath proper has to be duly noted. These are the *Kashi Vidvat Parishad* and the *Akhila Bharatiya Dharmasangha*.

It is agreed on all hands that the role played by members of these two organizations was a key factor in reviving the Jyotirmath in the year 1941. However, once Swami Brahmananda Saraswati was accepted as the Jyotirmath Shankaracharya, and there were no serious disputes about it at the time, further activities of these other institutions with respect to succession issues could be construed as unnecessary interference. Santananda and his successors trace their claim to a will of Brahmananda, but the *Kashi Vidvat Parishad* and the *Akhila Bharatiya Dharmasangha* have not accepted this. Interestingly, if the opinions of these bodies are to be set aside at any time after 1941, only the lineage of Vasudevananda (through Santananda) can be traced directly to Brahmananda, without any interruptions. However, it is precisely this lineage that seems to lack the necessary traditional support and that also seems to have now lost its previous legal standing.

Clearly, the role played by the *Kashi Vidvat Parishad* and the *Akhila Bharatiya Dharmasangha* was not questioned when Krishnabodha Ashrama was appointed to the Jyotirmath title, except perhaps by Santananda and his followers. As for the other Mathas and their heads, whatever their reservations may have been about Santananda's qualifications for the Shankaracharya title, their endorsement of Krishnabodha Ashrama's appointment and/or his subsequent appointment of Swarupananda meant that they tacitly approved of and accepted the actions of these two organizations.

Thus, both bodies got more firmly entrenched in matters relating to succession at Jyotirmath. Both Swarupananda Saraswati and Madhava Ashrama trace their claim to the Jyotirmath title through Krishnabodha Ashrama and *via* an acknowledgement of the right of the *Kashi Vidvat Parishad* and the *Akhila Bharatiya Dharmasangha* to decide the question of who would be the Shankaracharya of Jyotirmath. However, it could also be argued that these organizations have no right to unseat a Shankaracharya who is generally seen to be qualified for the post. Swarupananda was nominated by Krishnabodha

Ashrama, but these two organizations seem to have had no objections against him, at least till 1982, when he took charge of Dwaraka also.

The decade-long gap between Swarupananda's accession to the Dwaraka title and Madhava Ashrama's appointment in 1993/4 also seems significant. If Madhava Ashrama's claim of having been appointed by these two bodies is correct, this only means that the *Kashi Vidvat Parishad* and the *Akhila Bharatiya Dharmasangha* have tried to exercise a right over Jyotirmath for the third time. In 1941, the rationale was the revival of an extinct institution. In the 1950's, the rationale was the controversy over Brahmananda's will. In the 1990's, it has been the confusion arising due to Swarupananda's acceptance of the Shankaracharya title at the Dwaraka seat.

As Madhava Ashrama's Jyotirmath Shankaracharya title is very much dependent upon the latest intervention by these two bodies, it is not too far off the mark to say that his claim to the title holds only so long as he enjoys their confidence. It is not at all inconceivable that a fourth competing lineage will be put forth in the future, if either Madhava Ashrama or his successor develops differences of opinion with the leaders of one of these organizations. Another possible scenario is that Madhava Ashrama could stand to lose, if either Swarupananda or Vasudevananda manages to win their support.

The major lawsuits involved in the Jyotirmath succession are still pending, either unresolved in the lower courts, or under appeal in the Allahabad High Court. The property of Jyotirmath is presently divided between Santananda's (now Vasudevananda's) camp and Swarupananda's camp. On the way to Badrinath, in the village of Joshimath, there are two separate establishments, respectively controlled by Vasudevananda and Swarupananda.

The two are quite near each other, each calling itself *the* official Jyotirmath. I am unaware if Madhava Ashrama and his supporters have filed any lawsuits against either Swarupananda or Vasudevananda or both. My personal opinion is that it would be extremely difficult to legally uphold a claim that one man cannot be the head of two different Mathas. Unless the permanent legal standing of the *Kashi Vidvat Parishad* and the *Akhila Bharatiya Dharmasangha* in matters of succession at Jyotirmath can be proved, Madhava Ashrama's claim seems to be the weakest, from all possible perspectives. It is also unclear whether Madhava Ashrama and his followers have been able to get established in the old Matha property.

However, in this dispute, possession of or jurisdiction over the Matha's property is not the criterion which decides the identity of the legitimate Shankaracharya. The primary issue is one of authenticity and/or authority of the disciple lineage. As may be expected, no side is willing to concede the Shankaracharya title to one of the other two. Both Swarupananda Saraswati and Madhava Ashrama seriously deny the authority of the entire lineage beginning with Santananda Saraswati.

However, Madhava Ashrama and Swarupananda Saraswati also question the legitimacy of each other, and Vasudevananda Saraswati denies the rights of both. No quick resolution of the dispute seems possible, but none of these three are very young in age. Unless there is a consensus candidate acceptable to all parties in the current dispute, reconciliation seems quite impossible. When any one of them passes away, or nominates another person to his own title while alive, he will simply transfer to the person of his successor all the associated disputes against the other two (and *their* respective successors).

In this century, there have been similar disputes and lawsuits in connection with the succession at Puri and Dwaraka also. In contrast, south Indian Mathas in the Advaita Vedanta tradition are relatively non-controversial in this regard. To begin with, the southern Shankaracharyas are very selective about whom they personally initiate as monks, and tend not to have a large retinue of Sannyasin disciples with them. Traditionally, at Sringeri, the current Shankaracharya carefully chooses and trains his successor, who therefore is in the public eye for at least a few years before assuming charge.

The senior Acharya also typically withdraws from public life, and lets the chosen successor handle administrative issues. Or else, just before his own passing, the Shankaracharya clearly names his unique choice of successor from among his own disciples. Kanchi and other southern Mathas also usually take similar caution about their succession. In cases of southern Mathas with a following hailing mainly from particular communities, but where the lineage has suddenly become extinct, *e.g.* the Shirali Chitrapura Matha associated with Saraswata Brahmanas in southern Karnataka, the succession is revived by other means, often with the help and support of other well respected Mathas. Puri, Dwaraka and Jyotirmath seem to operate by very different procedures. Matters of succession are left rather nebulous, or stated in a will, with a choice of candidates for the successor. More often than not, the will is contested, so that the succession soon becomes a highly charged political issue, and the lineages become more complicated than in the Wars of the Roses.

The Role of Royalty: The relationship of all these institutions with members of current and erstwhile royal houses continues to remain important. Madhava Ashrama's visits to Nepal are usually highlighted to buttress his claim. Reading the publications of his followers, one gets the distinct impression that the support of the King of Nepal is seen as a very important factor in legitimating Madhava Ashrama's claim to the Jyotirmath Shankaracharya title. Reciprocally, the King of Nepal is described as the "King of all Hindus in the world," presumably because he is the only legitimate Hindu monarch left today, even if his role within Nepal has been much reduced in recent times. The *Kashi Vidvat Parishad* is usually supported by the current "king of Varanasi," who is, I assume, also a leader of the *Akhila Bharatiya Dharmasangha*. The revival of Jyotirmath in 1941 was facilitated by the role played by the then significant Hindu princes in north India. Members of India's erstwhile royalty continue to be important in Hindu religious institutions, even if their roles have become largely ceremonial, *e.g.* the Gajapati royal family in the Puri Jagannatha temple and the Travancore royal family in the TiruVanantapuram PadmanabhaSwami temple.

As for the other Mathas, in the pre-1947 era, administrative issues were routinely referred to the local rulers. For example, there are well documented records of the interactions of the Sringeri administration with various ruling houses, from the Vijayanagara emperors, many centuries ago, to the Mysore Wodeyars and Hyderabad Nizams, just before Indian independence. A large part of the rise of the Kanchi Matha to importance can be traced to the relationship that successive *Mathadhipatis* have maintained with the Maratha rulers of Tanjavur, the British administrators of Madras Presidency and the rulers of independent India.

Perhaps, all this is in keeping with the nature of generic Hindu religion, as ascetics are supposed to be supported mainly by householders, and the rulers are expected to uphold *dharma*. However, although Shankara describes the ascetic as an *atya'sramin* (one who is beyond all *Ashramas*), and asks rhetorically, "*katham varanasami bhavet?*," the situation of the contemporary Shankaracharyas is considerably more complicated. Ideally, a Mathadhipati pursues only other-worldly concerns, as he is first and foremost a monk.

In the Advaita tradition, he is also eventually expected to transcend all concerns of both this world and the next. However, a Matha is a social institution, and has one foot firmly planted in this world. The coronation of a Shankaracharya as the head of a Matha is called a *Pattabhisheka*, and closely resembles that of a classical Hindu king. Therefore, when the very issue of who is the legitimate Mathadhipati is confused or controversial, the realities of this world take centre-stage. Certainly, ever since Swami Vidyaranya's celebrated involvement with the Vijayanagara empire in the 14th century, and possibly from even earlier times, the Shankaracharyas and their Mathas have not been strangers to rulers of India and their politics.

It could be argued that the leaders of the Dasanami Sannyasin tradition have not woken up to the constitutional reality of democratic India, and that they continue to expect outmoded patterns of social behaviour, and that they continue to give undue importance to old royal dynasties. On the other hand, it could also be argued that this is simply one manifestation of a larger reality in India, whose new rulers have succeeded only in establishing modern political dynasties in place of the old royal ones. In any case, the current interactions among politicians, erstwhile royal houses and Sannyasins, as also disputes over which Sannyasin is a legitimate Shankaracharya, can be understood only along the lines of age-old *brahma-kshatra* relationships.

A Review of Cenkner's Work: In connection with the Shankaracharya tradition, Prof. Harzer-Clear mentions William Cenkner's book "A Tradition of Teachers", published by Motilal Banarsidass (1983). While Cenkner makes some valid observations about the inherent strengths and weaknesses of the Shankaracharya tradition, he is simply unreliable when it comes to factual details.

What his introduction calls the "inclusive tradition" of the Kanchi Matha is in fact a highly contested claim within the Dasanami tradition, with the potential to cause a deep schism. As for the Jyotirmath succession controversy, it has always been a matter of public knowledge in India, both because of the close physical proximity of the two separate establishments at Joshimath and because of the fact that all the people involved are public figures. Cenkner is astoundingly silent about it, and if he were to be believed, there is just no succession dispute at all, so that Santananda's position at Jyotirmath is unquestioned. He also specifically mentions the succession disputes at Dwaraka and Puri in this century, but strangely enough, he mentions a controversy where none existed, namely at Sringeri. Cenkner tells us that Niranjanadeva *Tirtha*'s coronation as the Shankaracharya of Puri in 1964 was conducted in the presence of the Shankaracharya of Jyotirmath, but neglects to inform us that this was Krishnabodha Ashrama, not Santananda.

Cenkner is also aware of the 1979 conference of the four Shankaracharyas, but seems unaware that it was Swarupananda, not Santananda or VishnudeVananda, who was described as the Shankaracharya of Jyotirmath then. Cenkner's 1995 edition seems to be a reprint of the 1983 edition, and although he has provided a fresh preface, which mentions Swarupananda Saraswati's name, he still does not mention the

Jyotirmath succession dispute. One is left with the impression that even as late as 1995, Santananda's position as Shankaracharya of Jyotirmath remained undisputed. Cenkner's silence regarding this dispute is extremely surprising, and he is either astoundingly naive or one has to suspect an unbecoming bias on his part.

Cenkner goes on to describe Santananda Saraswati as the head of all Dandi Sannyasins. This is extremely misleading, to say the least. There is no one head of all these Sannyasins. The heads of the Puri Matha have traditionally been Dandi Sannyasins, so that the Puri Shankaracharya might as well be described as their head. The Dwaraka and Sringeri Shankaracharyas, who are Paramahamsa Sannyasins, nominally the highest rank among the Dasanamis, would each have a better claim to being the overall head.

Moreover, in spite of the affiliation of all *Dasanami Sannyasins* with the four Mathas at Sringeri, Puri, Dwaraka and Jyotirmath, the Dasanami *Akhadas* are more important throughout north India, with numerous Mahamandaleshwaras, each with his own significant following. Even if there were no dispute about the Jyotirmath succession, Santananda's description as the head of all Dandi Sannyasins would be rather inaccurate. And the endorsement of Swarupananda by the Puri and Sringeri Mathas would make Santananda's claim (and now Vasudevananda's claim) even more tenuous. It may be added that acrimonious disputes about succession within one Matha and about the relative importance of the various Mathas are not limited to recent history. There are records of such disputes dating back to the 17th century and earlier. Often, such disputes are also accompanied by philosophical and doctrinal differences, so that a study of these would offer significant clues to the development of the post-Shankaran traditions of *Dasanami Sannyasins* and Advaita Vedanta. Cenkner seems completely blind to such crucial features of the Shankaracharya tradition, and this is because his historical research leaves much to be desired.

Appendix: Since the above discussion has repeatedly mentioned the names of some of the other Shankaracharyas in India, here is some information about their lineages. Only the 20th century is covered. Dates given here are dates of accession to the title, not the date of initiation into Sannyasa. It should be noted that there were succession controversies in both Puri and Dwaraka Mathas in the 19th century.

Puri (Purvamnaya)

- Shankara Madhusudana *Tirtha* (up to 1925)
- Bharati Krishna *Tirtha* (1925-1960)-He was the *Dwaraka* Shankaracharya till 1925. In 1921, Bharati Krishna *Tirtha* was prosecuted by the British Indian Government, for his involvement in the freedom struggle and his support of the Khilafat movement. He is also the author of a book called *Vedic Mathematics* and is the only major Shankaracharya to have travelled outside the subcontinent.
- Yogeswarananda *Tirtha* (1960-1961)-He passed away shortly after assuming charge, and without nominating a successor. He also wrote a book called *Mantra-Sastra*, published posthumously. His name is not included in the current list. Period of litigation-1961-1964. See the *Dwaraka* list below.
- Niranjanadeva *Tirtha* (1964-1992)
- Nischalananda Saraswati (since 1992)-Niranjanadeva stepped down in 1992.

Sringeri (Dakshinamnaya)

- Sacchidananda Sivabhinava Narasimha Bharati (up to 1912)-He is credited with having rediscovered the village of Kalati in Kerala, the birthplace of Shankara, and with instituting the current practice of observing the *Shankara Jayanti* celebrations.
- Chandrasekhara Bharati (1912-1954)-Author of a commentary on *Vivekachudamani*. An English translation has been published from the Bharatiya Vidya Bhavan.
- Abhinava Vidya *Tirtha* (1954-1989)
- Bharati *Tirtha* (since 1989)

Dwaraka (Pascimamnaya)

- Trivikrama *Tirtha* (until 1921)
- Bharati Krishna *Tirtha* (1921-1925)-See entry under *Puri* above.
- Swarupananda *Tirtha* (1925-??)
- Yogeswarananda *Tirtha* (??-1945)-Also see the *Puri* list above.
- Abhinava Sacchidananda *Tirtha* (1945-1982)-He played a major role in succession issues at *Puri* and **Jyotirmath**. He also established the *Adi Shankara Kaivalya Dhama* near Kedarnath.
- Swarupananda Saraswati (since 1982)

The Amnaya Mathas

The four *Mathas* established by Shankara are known in the tradition as the *amnaya Mathas*. Shankara is said to have assigned one of the four vedas to each of these *Mathas*, and the Acharyas and panditas of these four *Mathas* continue the tradition to this day. Accordingly, the Puri Matha is associated with the Rig veda, Sringeri with Yajurveda, Dvaraka with sama veda and Jyotirmath with atharva veda.

The ten Dasanami suffixes are distributed among these four *Mathas*-according to most traditions, *puri, Bharati* and *Sarasvati* with Sringeri; *Tirtha* and *ashrama* with Dvaraka; *Sagara*, *Parvata* and *Giri* with Jyotirmath, and *Vana* and *Aranya* with Puri. Many notable post-Shankaran authors, including Suresvara, Jnanaghana, Jnanottama, Anandagiri, Bharati *Tirtha*, Vidyaranya and others, can be found among the heads of these *Mathas*.

Of these four, Sringeri is the only institution that has had an unbroken line of succession from Shankara. Among the other three *Mathas*, the succession has been interrupted at one time or the other, for a variety of historical reasons. The longest hiatus in the line of succession was in the case of Jyotirmath, where the seat lay vacant for around 165 years. In the recent past, the Sringeri *Matha* has been involved, directly or indirectly, in stabilizing the line of succession in the other three *Mathas*.

From L to R: Sri Svarupananda Sarasvati (Jyotirmath), Sri Abhinava Vidya *Tirtha* (Sringeri), Sri Niranjana Deva *Tirtha* (Puri), Sri Abhinava Sacchidananda *Tirtha* (Dvaraka)-Meeting at Sringeri in 1979.

The successor to the title in a *Matha* is usually nominated by the presiding Shankaracharya of that *Matha*. It is quite normal to see Shankaracharyas who have become *Sannyasis* directly from the student life, without ever having been *grihasthas*. This is especially the norm in the Sringeri lineage. Thus, a Shankaracharya can be a very young man, sometimes barely out of his teens, when he takes charge at his *Mathas*. On the other hand, the Puri lineage has seen many heads who have become *Sannyasins* quite late in their lives, after passing through the *Grihastha* stage.

In cases where a Shankaracharya passes away without nominating a successor, or if there is a dispute about the succession, the head of one of the other *Mathas* is consulted to resolve the issue. Within this century itself, there have been instances where the Shankaracharyas of Sringeri, Dvaraka, and Puri have been called upon to resolve succession issues in one of the other *Mathas*. The Sringeri lineage names thirty-six successors to the Shankaracharya title, while Dvaraka has about seventy. The Puri list of Shankaracharyas has more than 140 names to date. The larger number of names in these two lists is probably because many of the presiding Shankaracharyas have been former *grihasthas*, who took charge at a comparatively older age and consequently held charge for shorter periods. The line of the Jyotirmath has many gaps in it, an unfortunate circumstance of history.

The position of the Shankaracharyas in modern Hinduism has often (quite wrongly) been compared to that of the Pope in Roman Catholicism. The four Shankaracharyas do not issue catechisms for all Hindus, nor do they claim sole right to decide on doctrinal issues. *Sri-mukhams* issued by the *Mathas* are very different in nature from papal bulls or encyclicals, and unlike the Vatican City, the four *Mathas* do not enjoy sovereign status. Rather, they are governed by the federal and state laws on religious and charitable trusts and endowments in independent India, and are often answerable to governmental bodies.

However, this should not be construed to mean that the Shankaracharyas are insignificant or that their importance is overrated. They are held in high respect by almost all sections of Hindus, but they also tend to get blamed by the modern media, somewhat unfairly, for everything that goes wrong in Hindu society! For all that, however, the Shankaracharyas generally lead quiet, secluded lives, as befits monks, and tend to avoid media attention. There are, of course, exceptions to this norm, and recent developments in India, especially the Ramjanmabhoomi-Babri Masjid issue, have forced all of them to take more active roles in public life.

Recent History of the Four Amnaya Mathas

Sringeri

Sri "Ugra" Narasimha Bharati (1817-1878) was well-known throughout India as a very saintly personality. He travelled far and wide, and had disciples all over India and even in Sri Lanka. He was succeeded by Sri Sacchidananda Sivabhinava Narasimha Bharati (1878-1912), who rediscovered Kaladi, Adi Shankaracharyas birth-place, and instituted Shankara Jayanti celebrations all over India. He also arranged for the publication of a comprehensive collection of Shankara's works, and initiated the practice of having the various Shankaracharyas meet for informal discussion and decision making. Following his lead, meetings took place at Kaladi, Hardwar, Prayag, etc. His successor, Sri Chandrasekhara Bharati (1912 to 1954), was an acclaimed *jivan-mukta.*

He wrote a commentary to Shankara's *Viveka-chudamami*. The first meeting of all four Shankaracharyas (*chaturamnaya-sammelanam)* in the 1200 year old tradition of post-Shankaran Advaita, took place at Sringeri, in 1979, under the leadership of Sri Abhinava Vidya *Tirtha* (1954-1989). Sri Bharati *Tirtha,* the presiding Shankaracharya of Sringeri, succeeded to the title in 1989. The Shankaracharyas of the four *amnaya Mathas* and the head of the Kanchi *Matha* held another conference at

Sringeri in 1993, following the events of December 1992 at Ayodhya, to express their concern at the politicization of religious issues, and resolved to lead a non-political effort to solve the Ramjanmabhoomi-Babri Masjid issue amicably.

Dvaraka

Sri Trivikrama *Tirtha* was the head of the Dvaraka *Matha* till the year 1921. He was succeeded by Sri Bharati Krishna *Tirtha*, who had a very interesting career. Beginning as a student of Vedanta at Sringeri, he became a Sannyasin under Sri Trivikrama *Tirtha* of Dvaraka, and succeeded to the Shankaracharya post at Dvaraka, in 1921. Soon after the first world war, he was prosecuted along with the Ali brothers and other Muslim leaders, by the colonial British government for treason, in connection with his involvement in the Indian Independence movement, and his support of the Khilafat movement. He is also said to have discovered some ancient *sutras* of basic arithmetic, which have been published as a book, under the title "Vedic mathematics". He was asked to take over the Puri *Matha* in 1925, when that seat fell vacant.

Accordingly, Sri Svarupananda *Tirtha* and Sri Yogesvarananda *Tirtha* followed at the Dvaraka seat. In the year 1945, Sri Abhinava Sacchidananda *Tirtha* was nominated as the Shankaracharya of Dvaraka, with Sri Bharati Krishna *Tirtha* performing the installation ceremonies. Before taking over at Dvaraka, Sri Abhinava Sacchidananda *Tirtha* was the head of the Mulabagal *Matha* in Karnataka. This was an old branch of the Dvaraka *Matha*, established in the 17th century, and with his appointment to the Dvaraka seat, the collateral lineage of Mulabagal *Matha* was merged with that of Dvaraka.

In later years, he was called upon to mediate the succession issues at both Puri and Jyotirmath. He also renovated the *Samadhi* site of *adi* Shankara at Kedarnath with assistance from the government of Uttar Pradesh. He passed away in 1982, following which Sri Svarupananda Sarasvati of Jyotirmath assumed charge at Dvaraka. Sri Abhinava Vidya *Tirtha* of Sringeri consecrated his appointment, and Sri Svarupananda has held dual charge at both Dvaraka and Jyotirmath since then.

Puri

This *Matha* is historically connected with the famous Jagannatha temple in Puri. It is also called the Govardhana *Matha*, and has an important branch in Puri itself, called the Shankarananda *Matha*. In the beginning of the century, the head was Sri Shankara Madhusudana *Tirtha*. Sri Bharati Krishna *Tirtha*, who was then at Dvaraka, took over as the Shankaracharya of Puri in 1925. Sri Bharati Krishna *Tirtha* visited the USA in the 1950's, at the invitation of the Self-Realization Fellowship. During this time, Sri Shankara Purusottama *Tirtha* supervised the Puri *Matha* on his behalf. After Sri Bharati Krishna *Tirtha* passed away in 1960, he was succeeded by Sri Yogesvarananda *Tirtha*, whose period was quite short, as he passed away in 1961. This lead to a brief period of uncertainty during which the succession at the *Matha* was being litigated. In 1964, Sri Niranjana Deva *Tirtha*, who was one of the nominees named in Sri Bharati Krishna *Tirtha*os will, was consecrated at the Puri seat by Sri Abhinava Sacchidananda *Tirtha* of Dvaraka. Sri Niranjana Deva *Tirtha* is known for his unpopular political views on volatile issues affecting Hindu people, like sati and cow protection. In 1992, he stepped down after nominating Sri Nischalananda Sarasvati as his successor, who is currently in charge at Puri.

Jyotirmath: Also known as Joshimath, it is located near Badrinath in the Himalayas, because of which it is also known as the Badrinath *Matha*. After a long hiatus of 165 years, this *Matha* was revived in the year 1941, under Sri Brahmananda Sarasvati, a disciple of Sri Krishnananda Sarasvati, who was originally from Sringeri. The appointment was made by a committee of pundits from Varanasi, and Sri Brahmananda's accomplishments helped re-establish the Jyotirmath as an important centre of traditional Advaita teaching in northern India. When he passed away in 1953, Sri Santananda Sarasvati succeeded him at this seat, according to the terms of a will. However, there was a dispute regarding the capacity of Sri

Santananda for the title and also about the validity of this will. This resulted in a major controversy that remains unresolved.

Karapatri Swami (Hariharananda Sarasvati), a well-known disciple of Sri Brahmananda, was asked to take over the Jyotirmath title, but he declined. To resolve the dispute, another committee of pundits from Varanasi was formed, under the guidance of Karapatri Swami and Sri Abhinava Sacchidananda *Tirtha* of Dvaraka. Sri Krishnabodhashrama was appointed as the new head of the *Matha*. When he passed away in the early 1970's, he nominated Sri Svarupananda Sarasvati, another disciple of Sri Brahmananda, as his successor. Sri Svarupananda continues as the Shankaracharya of Jyotirmath, and has also been in charge of Dvaraka since 1982.

Some people consider the rightful succession of the Jyotirmath title to be along the disciple line of Sri Santananda Sarasvati. He is said to have retired in 1980, in favor of his disciple, Sri VisnudeVananda Sarasvati, who has since passed away. Sri Santananda also passed away in December 1997, and has been succeeded by Sri Vasudevananda Sarasvati. Thus, there are at least two separate lineages at Jyotirmath currently, although it is Sri Svarupananda Sarasvati who is endorsed by the other *Amnaya Mathas*.

There is a third ascetic, named Sri Madhavashrama, who is another claimant to the Jyotirmath title, who contests both the claims of Sri Svarupananda and Sri Vasudevananda. Sri Madhavashrama is a disciple of Sri Krishnabodhashrama, who was nominated to the Jyotirmath title in the 1960's. His contention is that Sri Svarupananda cannot be accepted as the head of two different *amnaya Mathas* (Dvaraka and Jyotirmath), so that the Jyotirmath title has to revert to another disciple of Sri Krishnabodhashrama. According to publications supporting his claim, he was anointed in 1993 or 1994, under the guidance of Sri Niranjana Deva *Tirtha*, the former Shankaracharya of Puri. Thus, the dispute between two parties for the title of Jyotirmath Shankaracharya has now become a dispute among three different parties.

Other Mathas: Other than the four *amnaya Mathas*, there are a number of well-known *Mathas* owing allegiance to *Advaita* and the Shankaracharya lineage. Many of them were originally branches of one of the four *amnaya Mathas*, established officially by the parent *Matha*, and which grew into more or less independent institutions over time. Notable among these are the branch *Mathas* at Kumbhakonam, Sankhesvar, Kudali, Virupaksha (Hampi), Kolhapur (Karavir pitham), Sivaganga, Sakatapuram, etc.

In recent times, the *Matha* at Kanchipuram has been very active. Sometimes, notable *Sannyasis* of the *dasanami* order start their own *Mathas*, to cater to the spiritual needs of their followers. An example is the famous Upanisad Brahmendra *Matha* at Kanchipuram, which was founded in the 18th century by Ramachandrendra Sarasvati. Sometimes, succession controversies (as in the present Jyotirmath) also leads to the establishment of separate *Mathas*. A few *Mathas* of the *Nambudiri* community in Kerala also trace their foundation to Shankara himself, as do the sumeru and *Paduka Mathas* in Varanasi. However, the Kavale *Matha* of the *Gauda sarasvata* community in Goa traces its origin in 740 CE not to Shankaracharya, but through another disciple of Govinda Bhagavatpada.

In general, the various *Mathas* in India operate quite independent of one another. The Shankaracharya of the four original *Mathas* do not normally interfere with one another, nor do they seek to exercise any control, administrative or spiritual, on any of the other *Advaita Mathas* in India, unless specifically requested to do so. Although their heads are *Sannyasis* who lead completely detached lives, the *Advaita Mathas* are not immune to contemporary social and political pressures. Some *Mathas* deal with these pressures better than others. Manifestations of these pressures can be seen in the sometimes acrimonious rivalries between followers of two different *Mathas*, as also in the recurrent succession disputes in some of them. Such succession disputes sometimes lead to protracted litigation and the establishment of independent *Mathas* elsewhere.

Modern Institutions: In addition to the more traditional *Advaita Mathas* and *akhadas*, various *Sannyasis* of the *dasa-*

nami order have established some of the more well-known modern institutions, like the Ramakrishna Math and Mission (Swami Vivekananda), the Self-Realization Fellowship (Paramahamsa Yogananda), the Divine Life Society (Swami Shivananda), Yoga Vedanta Centre (Swami VishnudeVananda), the Chinmaya Mission (Swami Chinmayananda), and others. Among these, the founders of the Ramakrishna Mission, the Divine Life Society and the Chinmaya Mission trace their spiritual descent through the Sringeri *parampara*.

The Self-Realization Fellowship has links to the Puri *parampara*. These organizations usually teach some variant or the other of *Advaita-Vedanta*, generally combined with yoga practice, or an acceptance of the prophets of the Semitic religions, and/or an emphasis on social service. These modern institutions tend to have as much a presence in the West as in India, and their ideologies have come to be called by the generic name of neo-*Vedanta*. It remains to be seen whether these institutions will be the catalysts for the growth of a truly universal philosophy/religion that has been a dream of most of their founders.

There have been countless other nameless, realized masters over the centuries, who have realized the non-dual *Brahman*. As a living tradition of philosophy and religion, Advaita is not always restricted to *dasa-nami Sannyasis* in the lineage of Shankaracharya. For example, within the 20th century CE, one has the example of the famous mystic Sri Ramana Maharishi (1879-1950), who did not formally take *sannyasa*, but was nevertheless a *Jivaṇmukta*, who taught pure Advaita.

Bibliography

Ajit, M.: *Kundalini: The Arousal of the Inner Energy*, Destiny Books, New York, 1986.

Alain, A.: *Yoga: The Method of Re-Integration*, Johnson Publications, London, 1978.

Austin, B.: *Guide to Indian Philosophy*, G. K. Hall, Boston, 1989.

Benoytosh, B.: *Two Vajrayana Works*, Gaekwad's Oriental Series, Baroda, 1956.

Chand, Tara: *History of the Freedom Movement in India*, Kali Ram Press, Calcutta, 1984.

Chatterjee, A.K.: *A Comprehensive History of Vedanta*, Kali Ram Press, Calcutta, 1999.

Dandekar, R.N.: *Some Aspects of the History of Vedanta*, Radhey Shayam Press, Poona. 1989.

Datta, D.M.: *The Six Way of Knowing: A Critical Study of the Advaita Theory of Knowledge*, The University of Calcutta, Calcutta, 1994.

David, J.: *Shiva's Other Children: Religion and Social Identity amongst Overseas Indians,* Sage, Walnut Creek, 1998.

Deutsch Eliot: *Advaita Vedanta: A Philosophical Reconstruction*, The University Press of Hawaii, Honolulu, 1975.

Dutt, N.K.: *The Vedanta: Its Place as a System of Metaphysics*, Basu Mallik, Calcutta, 1985.

Farquhar, J.N.: *Modern Religious Movements in India*, Munshiram, New Delhi, 1978.

Flood, L.: *The Origins and Developments of Classical Hinduism* Beacon Press, Boston, 1998.

Gambhirananda, Swami: *Brahma Sutra Shamkar Bhasya,* Adavita Ashrama, Calcutta, 1986.

Gavin, F.: *An Introduction to Hinduism,* Cambridge University Press, England, 2001.

Gurumurthy, S.: *Hindu Heritage, Assimilative, Not Divisive.* Vigil, Madras 2001.

Harman, William: *The Sacred Marriage of a Hindu Goddess,* Indiana University Press, Bloomington, 1998.

Harold, W.: *The Swan's Wide Waters: Ramakrishna and Western Culture,* Kennikat Press, Port Washington, 1983.

Indich, William M.: *Consciousness in Advaita Vedanta*, Motilal Banarsidass Private Ltd., Delhi, 2003.

Iyar, K.A. : *Vedanta: The Science of Reality*, Ganesh and Co., Mardas, 1952.

Kenneth, W: *The Religion of the Hindus,* The Ronald Press Company, New York, 1965.

Kim, K.: *Hinduism, A Very Short Introduction,* Oxford University Press, New York, 2003.

Kramrisch, Stella: *Manifestations of Shiva*, Philadelphia Museum of Art, Philadelphia, 1994.

Krishna, W.: *Srimad Bhagavad Gita Bhashya of Sri Sankaracarya*, Ramakrishna Math, Madras, 1995.

Lex, H.: *Mother of the Universe: Visions of the Goddess, Tantric Hymns of Enlightenment*, Quest Books, Wheaton, 1994.

Mahadevan, T.M.P.: *Gaudapaada: A Study of Early Advaita*, University of Madras, Madras, 1954.

Mahalingam, T.V.: *Administration and Social Life under Vijayanagar*, University of Madras, Madras, 1969.

Majumdar, R.C.: *The Vedic Age*, Oxford, London, 1977.

Malkani, G.R.: *A Study of Reality*, The Modern Book Stall, Bombay, 1952.

Merlin, G.R.: *Vedanta Epistemology*, The Indian Institute of Philosophy, Lamar, 1978.

Mervyn, S.: *Problem of Two Truth in Buddhism and Vedanta*, Boston, New York, 1973.

Mircea, E.: *Rites and Symbols of Initiation*, Harper Torchbook, New York, 1965.

Mohan, G.: *The Mahabharata*, Munshiram Manoharlal, Delhi 1997.

Omar, G.: *Tantra: The Yoga of Sex*, Avon Books, New York, 1996.

Pandit, M. P.: *Thoughts of a Shakta*, Ganesh and Co., Madras, 1965.

Pierre, C.: *A Chronicle of the Last Pagans*, Cambridge, USA, 1990.

Pramathanath, M.: *Introduction to Vedanta Philosophy,* Basu Mallik Press, Calcutta: 1975.

Pushpendra, K.: *Hindu Religion and Ethics,* Asian Publication Services, New Delhi, 1997.

Raju, P. T.: *The Philosophical Traditions of India*, George Allen and Unwin, London, 1978.

Ramamurthi, A.: *Advaitic Mysticism of Shankara*, Shanti Niketan, West Bengal, 1974.

Ranade. R. D.: *A Constructive Survey of the Upanisadic Philosophy,* Saraswati Press, Poona, 1984.

Sakthi, Sri. K.: *Serpent Power*, Y. Subbaraya Sharma, Bangalore, 1971.

Sarma, Ramavatara: *Vedantism,* Williams Press, Calcutta, 1979.

Sen, L.: *The Ramayana of Valmiki*, Munshiram Manoharlal, New Delhi, 1978.

Shraddhananda, Swami: *Hindu Sangathan, Saviour of the Dying Race*, MacMillan, Delhi 1976.

Stein, Burton: *Vijayanagara*, Cambridge University Press, Cambridge, 1989.

Sukla, K.: *Acarya Gaudapada aur Pracina Vedanta*, Nagarjuna Bauddha Samsthana, Gorakhpur, 1983.

Tarkalankara, Chandrakanta: *Hindu Philosophy,* Basu Mallik Press, Calcutta, 1982.

Tattvabhusan, Sitanath: *The Vedanta and its Relation to Modern Thought,* Kali Ram Press, Calcutta, 1981.

Tiwari, N.: *Dimensions of Renunciation in Advaita Vedanta,* Motilal Banarsidass, Delhi, 1977.

Vable, D.: *The Arya Samaj. Hindu without Hinduism.* Vikas Publ., Delhi, 1983.

Vasant, G.: *Mysterious Kundalini,* Taraporevala, Bombay, 1972.

Venkatarama, A.: *Advaita Vedanta, according to Sankara,* Asia Publishing House, New York, 1965.

Vihari, D.: *The Essentials of Advaitism, Suresvara's Naishkarmyasiddhi Explained in English,* Punjab Oriental Series, Lahore, 1933.

Visvantha, K.: *Essentials of Hinduism,* Narosa Pub. House, New Delhi, 1989.

Vivekananda, Swami: *Raja Yoga,* Ramakrishna-Vivekananda Center, New York, 1973.

Walters, D.: *The Path: Autobiography of a Western Yogi (Swami Kriyananda),* Ananda Publications, Nevada City, 1977.

Wiener, Nobert: *God and Golem,* The M. I. T. Press, Cambridge, 1964.

Yujoon, J.G.: *Hermeneutical Essays on Vedaantic Topics,* Motilal Banarsidass, Delhi, 1986.

Zaehner, C.: *Vedanta,* Oxford University Press, New York, 1966.

Zenith, A.: *The Religion and Philosophy of the Veda and Upanishads,* Harvard University Press, Cambridge, 1925.

Index

A

Acharya, 134, 142, 164, 235.
Advaita Parampara, 28.
Advaita Vedanta, 4, 6, 8, 26, 29, 30, 123, 124, 125, 126, 127, 128, 129, 131, 133, 163, 164, 223, 235, 239.
Agamas, 16, 124.
Agni, 39, 223.
Aranyakas, 7, 41.
Arithmetic, 244.
Ashramas, 136, 237.
Atheism, 56, 79.
Authority, 16, 41, 42, 110, 113, 114, 115, 121, 214, 231, 235.

B

Beliefs, 138.
Bhagavad Gita, 2, 26, 124, 134, 135, 198, 209.
Bhakti Movement, 138.
Bhakti Yoga, 31, 218.
Bhaskara, 15, 16, 27.
Bhaskaracharya, 124.
Bhedabheda, 8, 15, 16, 19, 124, 125, 126, 147, 157, 228.
Brahma Sutras, 2, 7, 8, 26, 31, 118, 134, 135.
Brahmana, 228.
Buddha, 2, 52, 55, 57, 60, 67, 90, 103, 154, 155, 169, 198.
Buddhism, 6, 7.
Buddhist, 8, 10, 22, 56, 58, 59, 60, 110, 113, 134.

C

Civilization, 198.
Company, 18, 72, 73, 172, 178.
Creation, 34, 35, 36, 45, 110, 114, 127, 128, 129, 130, 132, 136, 165, 191, 194, 202.
Criticisms, 17.
Culture, 21, 197, 200.

D

Department, 142.
Development, 3, 8, 10, 114, 232, 239.
Devotion, 8, 9, 11, 18, 19, 27, 124, 134, 136, 179, 201, 216, 218.
Dharma, 118, 122, 161, 163, 165, 166, 198, 223, 237.

E

Equation, 22, 131, 132.
Ethics, 84, 88, 113, 114, 136, 137.
Evolution, 52, 109, 117, 142.

F

Faith, 3, 9, 14, 18, 23, 56, 79, 84, 85, 86, 94, 135, 179, 184, 195, 201.
Family, 91, 142, 171, 172, 177, 178, 192, 206, 236.

G

God, 2, 3, 7, 8, 9, 13, 15, 16, 18, 19, 20, 21, 22, 23, 25, 26, 27, 40, 44, 51, 53, 55, 56, 57, 58, 59, 61, 64, 65, 66, 67, 68, 73, 75, 79, 81, 82, 85, 86, 87, 88, 89, 90, 95, 96, 99, 100, 101, 102, 103, 104, 105, 106, 109, 110, 112, 115, 116, 119, 125, 126, 129, 131, 136, 137, 138, 139, 140, 144, 149, 151, 157, 158, 161, 162, 165, 176, 178, 179, 186, 197, 199, 200, 201, 202, 203, 206, 208, 216, 217, 218, 219, 222.
Gravitation, 45, 67, 108, 109.
Guru, 28, 29, 43, 92, 121, 122, 151, 158, 167, 169, 179, 181, 197, 198, 200, 210, 225, 227, 230.

H

Hindu Philosophy, 133, 137.
Hindu Scriptures, 1.
Hinduism, 2, 5, 6, 8, 118, 119, 124, 139, 221, 226, 242.

I

Institutions, 3, 41, 72, 73, 177, 221, 223, 224, 226, 230, 231, 233, 236, 247, 248.

J

Judges, 231.
Justice, 73, 151.

K

Karma, 5, 9, 16, 19, 31, 35, 40, 41, 42, 43, 60, 63, 105, 119, 123, 176, 201, 218, 219.
Krishna, 9, 11, 20, 21, 27, 70, 72, 75, 77, 119, 149, 150, 154, 161, 164, 170, 171, 186, 191, 193, 205, 223, 240, 241, 244, 245.

L

Languages, 178, 186.
Laws, 2, 51, 55, 137, 242.
Leadership, 222, 243.
Liberation, 9, 16, 18, 20, 26, 27, 37, 38, 39, 63, 122, 126, 131, 132, 136, 155, 185, 196, 199, 218, 219.
Literature, 16, 29, 124, 142, 178, 185, 186, 190, 228.

M

Madhava, 222, 230, 231, 233, 234, 235, 236.
Mahabharata, 10, 19, 139, 186.
Mantra, 122, 240.
Manu, 54.
Mathematics, 240, 244.
Medicine, 83.
Meditation, 5, 6, 7, 13, 16, 18, 37, 38, 76, 95, 123, 155, 179, 195, 197, 203, 229.
Metaphysics, 28, 135, 136, 211.
Mimamsa, 5, 13, 28, 118, 121, 122, 123, 139.
Moksha, 9, 13, 14, 18, 22, 63, 122, 124, 126, 131, 132, 136, 157, 222.

N

Narada, 107, 179.
Nirvana, 56, 58, 70.
Nyaya, 13, 44, 121, 124, 139, 141.

O

Opinion, 43, 86, 123, 166, 225, 230, 234.

P

Performance, 16, 20, 123.
Philosophical Thought, 21, 143.
Physics, 6, 7, 211, 212.
Politics, 237.
Prayer, 201, 203.
Production, 129.
Project, 49.
Puranas, 99, 139, 149, 186.
Purohit, 142.
Purva Mimamsa, 5, 118, 121, 122, 123.

R

Ramayana, 197.
Relationship, 3, 4, 6, 35, 106, 124, 136, 165, 166, 177, 195, 204, 236, 237.
Research, 40, 239.
Righteousness, 161.
Rishis, 29, 178, 180.
Rituals, 5, 118, 122, 142.
Rulers, 77, 224, 236, 237.

S

Sacraments, 142, 144.
Samhita, 5, 40, 45, 61, 97, 100.
Samskaras, 145, 173, 174, 175.
Sanatana Dharma, 163, 166.
Sannyasa, 224, 228, 240, 248.
Satya, 133, 165, 198.
Shankaracharya, 3, 4, 26, 28, 29, 42, 124, 128, 129, 130, 131, 132, 221, 222, 223, 224, 225, 226, 227, 228, 229, 230, 231, 232, 233, 234, 235, 236, 237, 238, 239, 240, 242, 243, 244, 245, 246, 247, 248.
Shishya, 167.
Shiva, 28, 29, 124, 136, 137, 139.
Shrine, 11, 156.

Space, 14, 25, 26, 40, 63, 84, 199, 204, 210, 211.
Stotras, 178.
Structure, 163, 230.

T

Tattva, 138, 141, 188, 189, 190, 194, 195.
Technology, 144.
Temples, 4, 11, 101.
Theology, 121, 138, 139.
Trade, 101.
Traditions, 4, 9, 29, 119, 124, 139, 239, 241.
Trimurti, 136, 139.

U

Upanishads, 1, 2, 3, 4, 5, 7, 8, 9, 10, 16, 18, 19, 21, 25, 26, 39, 41, 42, 43, 44, 46, 76, 105, 106, 109, 110, 111, 119, 122, 123, 124, 127, 129, 132, 134, 135, 139, 141, 142, 148, 149, 157, 163, 165, 197, 198.

V

Vaishnavism, 118, 134, 135, 137.
Vallabha, 9, 16, 19, 20, 27, 119, 126, 139.
Vasishtha, 28, 29, 207.
Veda, 5, 26, 40, 74, 121, 123, 131, 166, 186, 241.
Vedanta Sutras, 6, 31, 118, 119.
Vedas, 1, 5, 7, 8, 26, 28, 29, 37, 40, 41, 54, 55, 59, 60, 62, 65, 69, 81, 91, 95, 99, 102, 113, 114, 118, 121, 122, 123, 133, 136, 139, 146, 163, 165, 166, 198, 241.
Vishishtadvaita, 8, 27, 28, 43, 119, 133, 134, 135, 136, 138.
Vishnu, 11, 21, 27, 28, 124, 125, 134, 136, 137, 138, 139, 140.

W

Weapons, 51.
Welfare, 80, 148, 184, 186.
Worship, 22, 40, 41, 50, 56, 75, 81, 88, 92, 93, 94, 95, 100, 101, 102, 105, 106, 138, 144, 145, 177, 198, 201.

Y

Yajnas, 145.
Yajurveda, 241.
Yoga, 9, 10, 11, 31, 33, 142, 158, 162, 163, 178, 179, 195, 196, 197, 200, 201, 203, 205, 206, 207, 209, 210, 213, 217, 218, 248.